ESSENTIALS OF
MEDICAL GENOMICS

CONTENTS

PREFACE

This is a book about medical genomics, a new field that is attempting to combine knowledge generated from the Human Genome Project (HGP) and analytic methods from bioinformatics with the practice of medicine. From my perspective as a research molecular biologist, genomics has emerged as a result of automated high-throughput technologies entering the molecular biology laboratory and of bioinformatics being used to process the data. However, from the perspective of the medical doctor, medical genomics can be understood as an expanded form of medical genetics that deals with lots of genes at once, rather than just one gene at a time. This book is relevant to all medical professionals because *all* disease has a genetic component when hereditary factors are taken into account, such as susceptibility and resistance, severity of symptoms, and reaction to drugs. The National Institutes of Health (NIH) defines medical genetics to include molecular medicine (genetic testing and gene therapy), inherited disorders, and the ethical legal and social implications of the use of genetics technologies in medicine.

The ultimate goal of genetic medicine is to learn how to prevent disease or to treat it with gene therapy or a drug developed specifically for the underlying defect. Other applications include pharmacogenomics and patient counseling about individual health risks, which will be facilitated by new DNA chip technology. Concerns include how to integrate genetic technology into clinical practice and how to prevent genetic-based discrimination.

Collins 1999

Before a coherent discussion of genomics is possible, it is necessary to define what is meant by a genome. A genome is the total set of genetic information present in an organism. Generally, every cell in an organism has a complete and identical copy of the genome, but there are many exceptions to this rule. Genomes come in different shapes and sizes for different types of organisms, although there is not always a simple and obvious connection between the size and complexity of an organism and its genome.

An operational definition of genomics might be: The application of high-throughput automated technologies to molecular biology. For the purposes of this book, genomics is defined broadly to include a variety of technologies, such as genome sequencing, DNA diagnostic testing, measurements of genetic variation and polymorphism, microarray gene expression, proteomics (measurements of all protein present in a cell or tissues), pharmacogenomics (genetic predictions of drug reactions), gene therapy, and other forms of DNA drugs. A philosophical definition of genomics might be: A holistic or systems approach to information flow within the cell.

Biology is complex. In fact, complexity is the hallmark of biological systems from cells to organisms to ecosystems. Rules have exceptions. Information tends to flow in branching feedback loops rather than in neat chains of cause and effect. Biological systems are not organized according to design principles that necessarily make sense to humans. Redundancy and seemingly unnecessary levels of interlocking dynamic regulation are common. Molecular biology is a profoundly reductionist discipline—complex biological systems are dissected by forcing them into a framework so that a single experimental variable is isolated. Genomics must embrace biological complexity and resist the human tendency to look for simple solutions and clear rules. Genomic medicine will not find a single gene for every disease. To successfully modify a complex dynamic system that has become unbalanced in a disease state will require a much greater subtlety of understanding than is typical in modern medicine.

The HGP was funded by the United States and other national governments for the express purpose of improving medicine. Now that the initial goals of the project have largely been met, the burden has shifted from DNA sequencing technologists to biomedical researchers and clinicians who can use this wealth of information to bring improved medicine to the patients—medical genomics. The initial results produced by these genome-enabled researchers give every indication that the promises made by those who initially proposed the genome project will be kept.

The initial sequencing of the 3.2 billion base pairs of the human genome is now essentially complete. A lot of fancy phrases have been used to tout the enormous significance of this achievement. Francis Collins, director of the National Human Genome Research Institute called it "a bold research program to characterize in ultimate detail the complete set of genetic instructions of the human being." President Clinton declared it "a milestone for humanity."

This book goes light on the hyperbole and the offering of rosy long-term predictions. Instead, it focuses on the most likely short-term changes that will be experienced in the practice of medicine. The time horizon here is 5 years into the future for technologies that are currently under intensive development and 10 years for those that I consider extremely likely to be implemented on a broad scale. In 5 years' time, you will need to throw this book away and get a new one to remain abreast of the new technologies coming over the horizon.

This book is an outgrowth of a medical genomics course that I developed in 2000 and 2001 as an elective course for medical students at the New York University School of Medicine. Based on this experience, I can predict with confidence that medical genomics will become an essential and required part of the medical school curriculum in 5 years or less. I also learned that medical students (and physicians in general) need to learn to integrate an immense amount of information, so they tend to focus on the essentials and they ask to be taught "only what I really need to know."

It is difficult to boil down medical genomics to a few hours' worth of bullet points on *PowerPoint* slides. There is a *lot* of background material that the student must keep in mind to understand the new developments fully. Medical genomics relies heavily on biochemistry, molecular biology, probability and statistics, and most of all on classical genetics.

My specialty is in the relatively new field of bioinformatics, which has recently come in from the extreme reaches of theoretical biology to suddenly play a key role in the interpretation of the human genome sequence for biomedical research. Bioinformatics is the use of computers to analyze biological information—primarily DNA and protein sequences. This is a useful perspective from which to observe and discuss the emerging field of medical genomics, which is based on the analysis and interpretation of biological information derived from DNA sequences. Two chapters were written by colleagues who are deep in the trenches of the battle to integrate genome technologies into the day-to-day practice of medicine in a busy hospital. Harry Ostrer is the director of the Human Genetics Program at the New York University Medical Center, where he overseas hundreds of weekly genetic tests of newborns, fetuses, and prospective parents. John Hay is co-director of the molecular biology core lab for the New York University General Clinical Research Center and the principle investigator of numerous projects to develop and test gene therapy methods.

Stuart M. Brown

REFERENCE

Collins F., Geriatrics 1999; 54: 41–47

*To Kim, who encourages me to write
and to Justin and Emma, who make me proud*

ACKNOWLEDGMENTS

This book grew out of a course that I taught to medical students at NYU School of Medicine in 2000 and 2001 as part of an interdisciplinary effort called the "Master Scholars Program." Joe Sanger, the Society Master for Informatics and Biotechnology, cajoled, coaxed, and guilt-tripped me into teaching the course. I also thank Ross Smith for hiring me as the *Molecular Biology Consultant* to the Academic Computing unit at NYU School of Medicine. He created a work environment where I could freely organize my time between teaching, consulting, maintaining the core computing systems, and writing. I must also thank my System Managers Tirza Doniger and Guoneng Zhong for picking up the slack for maintaining the UNIX systems and handling the tech support questions so that I could have time for writing.

In a larger context, I must thank my wife Kim for encouraging me to write something less technical that would appeal to wider audience, and for frequently suggesting that I take "writing days" to finish up the manuscript. She also provided some clutch help on several of the figures.

At Wiley, I thank Luna Han for having interest and faith in my concept for this book, and Kristin Cooke Fasano for sheparding me through all of the details that are required to make a manuscript into a book.

Finally, I must give credit to Apple Computer for the wonderful and light iBook that allowed me to do a great deal of the writing on the Long Island Railroad.

Stuart M. Brown

DECIPHERING THE HUMAN GENOME PROJECT

The Human Genome Project is a bold undertaking to under-stand, at a fundamental level, all of the genetic information required to build and maintain a human being. The human **genome** is the complete information content of the human cell. This information is encoded in approximately 3.2 billion base pairs of DNA contained on 46 **chromosomes** (22 pairs of auto-somes plus 2 sex chromosomes) (Fig. 1-1). The completion in 2001 of the first draft of the human genome sequence is only the first phase of this project (Lander et al., 2001; Venter et al., 2001). *This figure also appears in the Color Insert section.*

To use the metaphor of a book, the draft genome sequence gives biology all of the letters, in the correct order on the pages, but without the ability to recognize words, sentences, punctua-tion, or even an understanding of the language in which the book is written. The task of making sense of all of this raw biological information falls, at least initially, to **bioinformatics** specialists who make use of computers to find the words and decode the language. The next step is to integrate all of this information into a new form of experimental biology, known as

Essentials of Medical Genomics, Edited by Stuart M. Brown.
ISBN 0-471-21003-X. Copyright © 2003 by Wiley-Liss, Inc.

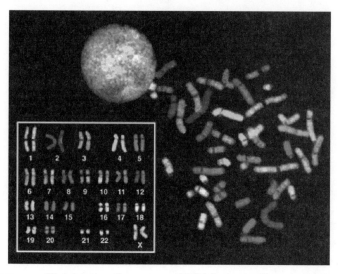

FIGURE 1-1. The human karyotype (SKY image). Figure also appears in Color Figure Section. Reprinted with permission from Thomas Ried National Cancer Institute.

genomics, that can ask meaningful questions about what is happening in complex systems where tens of thousands of different genes and proteins are interacting simultaneously.

The primary justification for the considerable amount of money spent on sequencing the human genome (from both governments and private corporations), is that this information will lead to dramatic medical advances. In fact, the first wave of new drugs and medical technologies derived from genome information is currently making its way through clinical trials and into the health-care system. However, in order for medical professionals to make effective use of these new advances, they need to understand something about genes and genomes. Just as it is important for physicians to understand how to Gram stain and evaluate a culture of bacteria, even if they never actually perform this test themselves in their medical practice, it is important to understand how DNA technologies work in order to appreciate their strengths, weaknesses, and peculiarities.

However, before we can discuss whole genomes and genomic technologies, it is necessary to understand the basics of how genes function to control biochemical processes within the cell (molecular biology) and how hereditary information is transmitted from one generation to the next (genetics).

THE PRINCIPLES OF INHERITANCE

The principles of genetics were first described by the monk Gregor Mendel in 1866 in his observations of the inheritance of traits in garden peas. Mendel described "differentiating characters" (*differierende Merkmale*) that may come in several forms. In his monastery garden, he made crosses between strains of garden peas that had different characters, each with two alternate forms that were easily observable, such as purple or white flower color, yellow or green seed color, smooth or wrinkled seed shape, and tall or short plant height. (These alternate forms are now known as **alleles**.) Then he studied the distribution of these forms in several generations of offspring from his crosses.

Mendel observed the same patterns of inheritance for each of these characters. Each strain, when bred with itself, showed no changes in any of the characters. In a cross between two strains that differ for a single character, such as pink vs. white flowers, the first generation of hybrid offspring (F_1) all looked like one parent—all pink. Mendel called this the **dominant** form of the character. After self-pollinating the F_1 plants, the second-generation plants (F_2) showed a mixture of the two parental forms (Fig. 1-2). This is known as **segregation.** The **recessive** form that was not seen in the F_1 generation (white flowers) was found in one-quarter of the F_2 plants.

Mendel also made crosses between strains of peas that differed for two or more traits. He found that each of the traits was assorted independently in the progeny—there was no connection between whether an F_2 plant had the dominant or recessive form for one character and what form it carried for another character (Fig. 1-3).

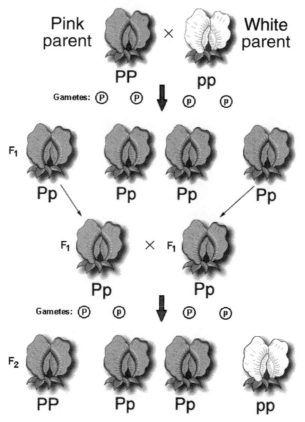

FIGURE 1-2. Mendel observed a single trait segregating over two generations.

Mendel created a theoretical model (now known as Mendel's Laws of Genetics) to explain his results. He proposed that each individual has two copies of the hereditary material for each character, which may determine different forms of that character. These two copies separate and are subjected to independent assortment during the formation of gametes (sex cells). When a new individual is created by the fusion of two sex cells, the two copies from the two parents combine to produce a visible trait, depending on which form is dominant and which is recessive. Mendel did not propose any physical explanation for

Mendel: dihybrid cross

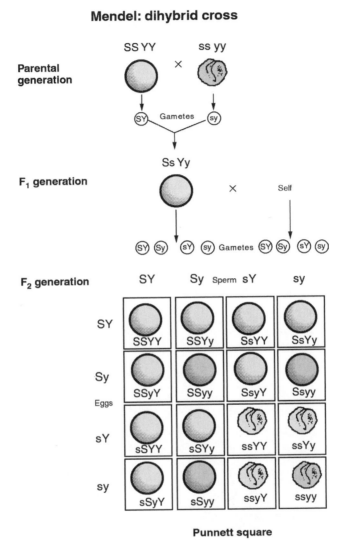

Punnett square

FIGURE 1-3. A cross in which two independent traits segregate.

how these traits were passed from parent to progeny; his characters were purely abstract units of heredity.

Modern genetics has completely embraced Mendel's model with some additional detail. There may be more than two different alleles for a gene in a population, but each individual has only two, which may be the same (**homozygous**) or different (**heterozygous**). In some cases, two different alleles combine to produce an intermediate form in heterozygous individuals; for example, a red flower allele and a white flower allele may combine to produce a pink flower; and in humans, a type A allele and a type B allele for red blood cell antigens combine to produce the AB blood type.

GENES ARE ON CHROMOSOMES

In 1902, Walter Sutton, a microscopist, proposed that Mendel's heritable characters resided on the chromosomes that he observed inside the cell nucleus (Fig. 1-4). Sutton noted that "the association of paternal and maternal chromosomes in pairs and their subsequent separation during cell division ... may constitute the physical basis of the Mendelian law of heredity" (Sutton, 1903).

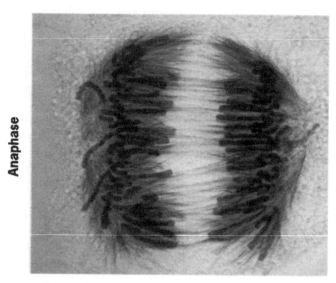

FIGURE 1-4. Chromosomes during anaphase in a lily cell.

In 1909, the Danish botanist Wilhelm Johanssen coined the term *gene* to describe Mendel's heritable characters. In 1910, Thomas Hunt Morgan (1910) found that a trait for white eye color was located on the X chromosome of the fruit fly and was inherited together with a factor that determines sex. A number of subsequent studies by Morgan and others showed that each gene for a particular trait was located at a specific spot, or **locus,** on a chromosome in all individuals of a species. The chromosome was a linear organization of genes, like beads on a string. Throughout the early part of the twentieth century, a gene was considered to be a single, fundamental, indivisible unit of heredity, in much the same way as an atom was considered to be the fundamental unit of matter.

Each individual has two copies of each chromosome, having received one copy from each parent. In sexual cell division (**meiosis**), the two copies of each chromosome in the parent are separated and randomly assorted among the sex cells (sperm or egg) in a process called segregation. When a sperm and an egg cell combine, a new individual is created with new combinations of alleles. It is possible to observe the segregation of chromosomes during meiosis using only a moderately powerful microscope. It is an aesthetically satisfying triumph of biology that this observed segregation of chromosomes in cells exactly corresponds to the segregation of traits that Mendel observed in his peas.

RECOMBINATION AND LINKAGE

In the early part of the twentieth century, Mendel's concepts of inherited characters were broadly adopted both by practical plant and animal breeders as well as by experimental geneticists. It rapidly became clear that Mendel's experiments represented an oversimplified view of inheritance. He must have intentionally chosen characters in his peas that were inherited independently. In breeding experiments in which many traits differ between parents, it is commonly observed that progeny inherit pairs or groups of traits together from one parent far more frequently than would be expected by chance alone. This observation fit nicely into the chromosome model of inheritance—if two genes are located on the same chromosome, then they will be inherited together when that chromosome segregates into a gamete and that gamete becomes part of a new individual.

However, it was also observed that "linked" genes do occasionally separate. A theory of **recombination** was developed to explain these events. It was proposed that during the process of meiosis the homologous chromosome pairs line up and exchange segments in a process called crossing-over. This theory was supported by microscopic evidence of X-shaped structures called chiasmata forming between paired homologous chromosomes in meiotic cells (Fig. 1-5).

If a parent cell contains two different alleles for two different linked genes, then after the cross-over, the chromosomes in the gametes will contain new combinations of these alleles. For example, if one chromosome contains alleles **A** and **B** for two genes, and the other chromosome contains alleles **a** and **b**, then—without cross-over—all progeny must inherit a chromosome from that parent with either an **A-B** or an **a-b** allele combination. If a cross-over occurs between the two genes, then the resulting chromosomes will contain the **A-b** and **a-B** allele combinations (Fig. 1-6).

FIGURE 1-5. Chiasmata visible in an electron micrograph of a meiotic chromosome pair.

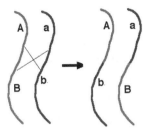

FIGURE 1-6. A single cross-over between a chromosome with **A-B** alleles and a chromosome with **a-b** alleles, forming **A-b** and **a-B** recombinant chromosomes.

Morgan, continuing his work with fruit flies, demonstrated that the chance of a cross-over occurring between any two linked genes is proportional to the distance between them on the chromosome. Therefore, by counting the frequency of cross-overs between the alleles of a number of pairs of genes, it is possible to map those genes on a chromosome. (Morgan was awarded the 1933 Nobel Prize in medicine for this work.) In fact, it is generally observed that on average, there is more than one cross-over between every pair of homologous chromosomes in every meiosis, so that two genes located on opposite ends of a chromosome do not appear to be linked at all. On the other hand, alleles of genes that are located very close together are rarely separated by recombination (Fig. 1-7).

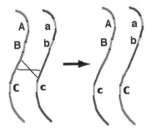

FIGURE 1-7. Genes A and B are tightly linked so that they are not separated by recombination, but gene C is farther away. After recombination occurs in some meiotic cells, gametes are produced with the following allele combinations: **A-B-C**, **a-b-c**, **A-B-c**, and **a-b-C**.

The relationship between the frequency of recombination between alleles and the distance between gene loci on a chromosome has been used to construct genetic maps for many different organisms, including humans. It has been a fundamental assumption of genetics for almost 100 years that recombinations occur randomly along the chromosome at any location, even within genes. However, recent data from DNA sequencing of genes in human populations suggest that there are recombination hot spots and regions where recombination almost never occurs. This creates groups of alleles from neighboring genes on a chromosome, known as **haplotypes,** that remain linked together across hundreds of generations.

GENES ENCODE PROTEINS

In 1941, Beadle and Tatum showed that a single mutation, caused by exposing the fungus *Neurospora crassa* to X-rays, destroyed the function of a single enzyme, which in turn interrupted a biochemical pathway at a specific step. This mutation segregated among the progeny exactly as did the traits in Mendel's peas. The X-ray damage to a specific region of one chromosome destroyed the instructions for the synthesis of a specific enzyme. Thus a gene is a spot on a chromosome that codes for a single enzyme. In subsequent years, a number of other researchers broadened this concept by showing that genes code for all types of proteins, not just enzymes, leading to the "One Gene, One Protein" model, which is the core of modern molecular biology. (Beadle and Tatum shared the 1958 Nobel Prize in medicine.)

GENES ARE MADE OF DNA

The next step in understanding the nature of the gene was to dissect the chemical structure of the chromosome. Crude biochemical purification had shown that chromosomes are composed of both protein and DNA. In 1944, Avery, MacLeod, and McCarty conducted their classic experiment on the "transforming principle." They found that DNA purified from a lethal S (smooth) form of *Streptococcus pneumoniae* could transform a harmless R (rough) strain into the S form (Fig. 1-8). Treatment of the DNA with protease to destroy all of the protein had no effect, but treatment with DNA-degrading enzymes blocked the transformation. Therefore, the information that transforms the bacteria from R to S must be contained in the DNA.

Hershey and Chase confirmed the role of DNA with their classic 1952 "blender experiment" on bacteriophage viruses. The phages were radioactively labeled with either ^{35}S in their proteins or ^{32}P in their DNA. The researchers used a blender to interrupt the process of infection of *Escherichia coli* bacteria by the phages. Then they separated the phages from the infected bacteria by centrifugation and collected the phages and bacteria separately. They observed that the ^{35}S-labeled protein remained with the phage while the ^{32}P-labeled DNA was found inside the infected bacteria (Fig. 1-9). This proved that it is the DNA portion of the virus that enters the bacteria and contains the

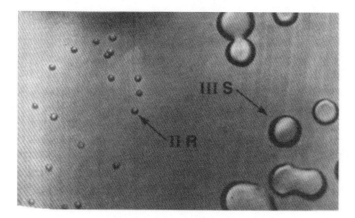

FIGURE 1-8. Rough and smooth *Streptococcus* cells.

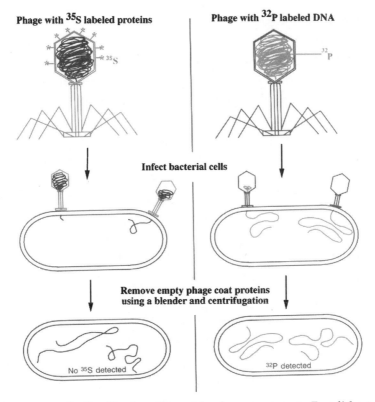

Phage with ³⁵S labeled proteins

Phage with ³²P labeled DNA

Infect bacterial cells

Remove empty phage coat proteins
using a blender and centrifugation

No ³⁵S detected

³²P detected

FIGURE 1-9. In the Hershey-Chase blender experiment, *E. coli* bacteria were infected with either ³⁵S-labeled proteins or ³²P-labeled DNA. After removing the phages, the ³²P-labeled DNA, but not the ³⁵S-labeled protein, was found inside the bacteria. Reprinted with permission from the DNA Science Book, CSHL Press.

genetic instructions for producing new phage, not the proteins, which remain outside. (Hershey was awarded the 1969 Nobel Prize for this work.)

DNA STRUCTURE

Now it was clear that genes are made of DNA, but how does this chemically simple molecule contain so much information? DNA is a long polymer molecule that contains a mixture of four different chemical subunits: adenine (A), cytosine (C), guanosine (G), and thymine (T). These subunits, known as nucleotide bases, have similar two-part chemical structures that contain a deoxyribose sugar and a nitrogen ring (Fig. 1-10), hence the name deoxyribonucleic acid. The real challenge was to understand how the nucleotides fit together in a way that can contain a lot of information.

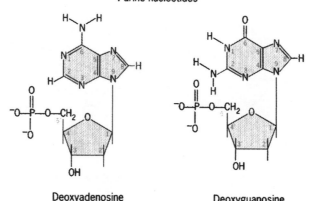

Pyrimidine nucleotides

Deoxythymidine
monophosphate, dTMP

Deoxycytidine
monophosphate, dCMP

Purine nucleotides

Deoxyadenosine
monophosphate, dAMP

Deoxyguanosine
monophosphate, dGMP

FIGURE 1-10. The DNA bases.

In 1950, Edwin Chargaff discovered that there was a consistent one-to-one ratio of adenine to thymine and of guanine to cytosine in any sample of DNA from any organism. In 1951, Linus Pauling and R. B. Corey described the α-helical structure of a protein. Shortly thereafter, Rosalind Franklin provided X-ray crystallographic images of DNA to James Watson and Francis Crick, which showed many similarities to the α-helix described by Pauling (Fig. 1-11). Watson and Crick's crucial insight was to realize that DNA formed a double helix with complementary bonds between adenine-thymine and guanine-cytosine pairs.

The Wastson-Crick model of the structure of DNA looks like a twisted ladder. The two sides of the ladder are formed by strong covalent bonds between the phosphate on the 5′ carbon of one deoxyribose sugar and the methyl side groups of the

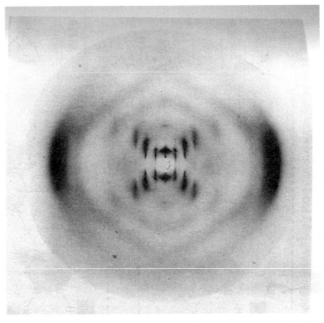

FIGURE 1-11. Franklin's X-ray diffraction picture of DNA.

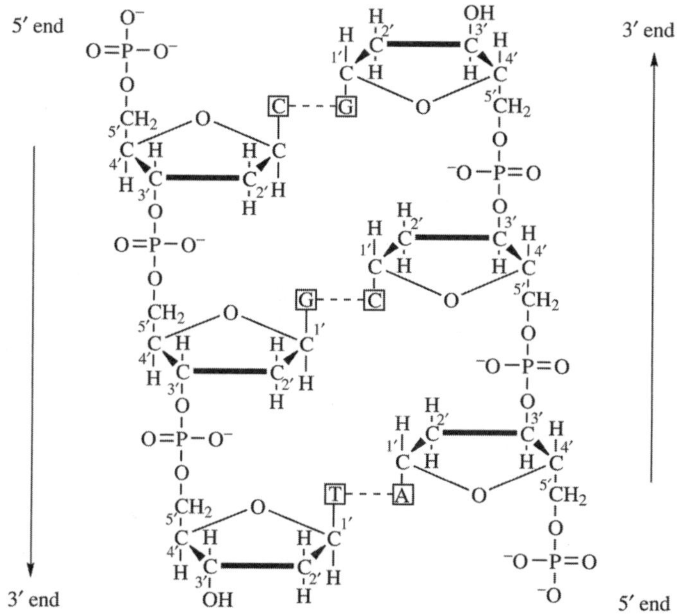

FIGURE 1-12. The DNA phosphate bonds. Reproduced, with permission, from T. Brown, *Genomes* 2nd edn. Copyright 2002, BIOS Scientific Publishers Ltd.

3' carbon of the next (a phosphodiester bond) (Fig. 1-12). Thus the deoxyribose sugar part of each nucleotide is bonded to the one above and below it, forming a chain that is the backbone of the DNA molecule. The phosphate to methyl linkage of the deoxyribose sugars give the DNA chain a direction, or polarity, generally referred to as 5' to 3'. Each DNA molecule contains two parallel chains that run in opposite directions and form the sides of the ladder.

The rungs of the ladder are formed by weaker hydrogen bonds between the nitrogen ring parts of pairs of the nucleotide bases (Fig. 1-13). There are only two types of base pair bonds: adenine bonds with thymine, and guanine bonds with cytosine. The order of nucleotide bases on the two sides of the ladder always reflects this complementary base pairing—so that wherever there is an A on one side, there is always a T on the other

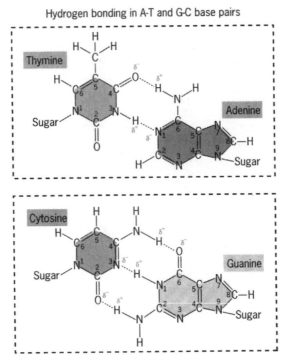

FIGURE 1-13. The DNA hydrogen bonds.

side, and vice versa. Since the A-T and G-C units always occur together, they are often referred to as **base pairs.** The G-C base pair has three hydrogen bonds, whereas the A-T pair has only two, so the bonds between the G-C bases are more stable at high temperatures than the A-T bonds. The nucleotide bases are strung together on the polydeoxyribose backbone like beads on a string. It is the particular order of the four different bases as they occur along the string that contains all of the genetic information.

Watson and Crick realized that this model of DNA structure contained many implications (Fig. 1-14). First, the two strands of

FIGURE 1-14. Watson and Crick demonstrate their model of the DNA double helix. Reprinted with permission, Photo Researchers, Inc.

the double helix are complementary. Thus, one strand can serve as a template for the synthesis of a new copy of the other strand—a T is added to the new strand wherever there is an A, a G for each C, etc.—perfectly retaining the information in the original double strand. In 1953, in a single-page paper in the journal *Nature,* Watson and Crick wrote, with a mastery of understatement:

> It has not escaped our attention that the specific pairing we have postulated immediately suggests a possible copying mechanism for the genetic material.

So, in one tidy theory, the chemical structure of DNA explains how genetic information is stored on the chromosome and how it is passed on when cells divide. That is why Watson and Crick won the 1962 Nobel Prize (shared with Maurice Wilkins).

If the two complementary strands of a DNA molecule are separated in the laboratory by boiling (known as denaturing the DNA), then they can find each other and pair back up, by reforming the complementary A-T and C-G hydrogen bonds (annealing). Bits of single-stranded DNA from different genes do not have perfectly complementary sequences, so they will not pair up in solution. This process of separating and reannealing-complementary pieces of DNA is known as DNA hybridization, and it is a fundamental principle behind many different molecular biology technologies. (see Chapter 2)

THE CENTRAL DOGMA

Crick followed up in 1957 with a theoretical framework for the flow of genetic information in biological systems. His theory, which has come to be known as the **Central Dogma** of molecular biology, is that DNA codes for genes in a strictly linear fashion— a series of DNA bases corresponds to a series of amino acids in a protein. DNA is copied into RNA, which serves as a template for protein synthesis. This leads to a nice neat conceptual diagram of the flow of genetic information within a cell: DNA is copied to more DNA in a process known as **replication,** and DNA is transcribed into RNA, which is then translated into protein (Fig. 1-15).

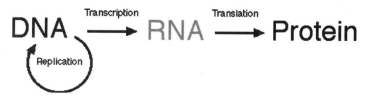

FIGURE 1-15. The Central Dogma of molecular biology, as described by Crick (1957).

DNA REPLICATION

Every ordinary cell (somatic cell) in an organism has a complete copy of that organism's genome. In mammals and other **diploid** organisms, that genome contains two copies of every chromosome, one from each parent. As an organism grows, cells divide by a process known as **mitosis.** Before a cell can divide, it must make a complete copy of its genome so that each daughter cell will receive a full set of chromosomes. All of the DNA is replicated by a process that makes use of the complementary nature of the base pairs in the double helix.

In DNA replication, the complementary base pairs of the two strands of the DNA helix partially separate and new copies of both strands are made simultaneously. A DNA polymerase enzyme attaches to the single-stranded DNA and synthesizes new strands by joining free DNA nucleotides into a growing chain that is exactly complementary to the template strand (Fig. 1-16). In addition to a template strand and free nucleotides, the DNA polymerase also requires a primer—a short piece of DNA that is complementary to the template. The primer binds to its complementary spot on the template to form the start of the new strand, which is then extended by the polymerase, adding one complementary base at a time, moving in the 5' to 3' direction. In natural DNA replication, the primer binds to specific spots on the chromosome known as the origins of replication.

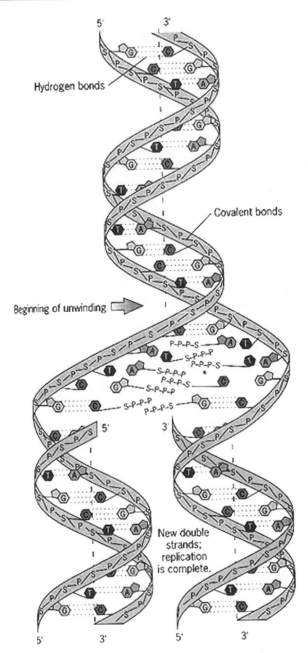

FIGURE 1-16. DNA replication showing the synthesis of two complementary strands at a replication fork.

This semiconservative replication process was demonstrated quite eloquently by the famous 1958 experiment of Meselson and Stahl. They grew bacteria in a solution of free DNA nucleotides that contained heavy ^{15}N atoms. After many generations, the bacterial DNA contained heavy atoms throughout. Then the bacteria were transferred to a growth medium that contained normal nucleotides. After one generation, all bacterial cells had DNA with half heavy and half light nitrogen atoms. After two generations, half of the bacteria had DNA with normal nitrogen and the other half had one heavy and one light DNA strand. After each cell division, both daughter cells have chromosomes made up of DNA molecules that have one strand from the parent cell and the other strand that was newly synthesized. This method of semiconservative DNA replication is common to all forms of life on earth from bacteria to humans.

This mechanism of DNA replication has been exploited in modern DNA sequencing biochemistry, which often uses DNA polymerase from bacteria or other organisms to copy human (or any other) DNA. Key aspects of the replication process to keep in mind are that the DNA is copied linearly, one base at a time, from a specific starting point (origin) that is matched by a short primer of complementary sequence. The primer is extended by the reaction as new nucleotides are added, so that the primer becomes part of the newly synthesized complementary strand.

TRANSCRIPTION

The DNA in the chromosomes contains genes, which are instructions for the manufacture of proteins, which in turn control all of the metabolic activities of the cell. For the cell to use these instructions, the genetic information must be moved from the chromosomes inside the nucleus out to the cytoplasm, where proteins are manufactured. This information transfer is done using messenger RNA (mRNA) as an intermediary molecule. RNA (ribonucleic acid) is a polymer of nucleotides and is chemically similar to DNA, but with several distinct differences. First, RNA is a single-stranded molecule, so it does not form a double helix. Second, RNA nucleotides contain ribose rather than deoxyribose sugars. Third, RNA uses uracil (U) in place of thymine, so the common abbreviations for the RNA bases are A, U, G, and C. As a result of these chemical differences, RNA is much less stable in the cell. In fact, the average RNA molecule has a life span that can be measured in minutes, whereas DNA can be recovered from biological materials that are many thousands of years old.

The process of transcription of DNA into mRNA is similar to DNA replication. A region of the double helix is separated into two strands. One of the single strands of DNA (the coding strand) is copied, one base at a time, into a complementary strand of RNA. The enzyme RNA polymerase catalyzes the incorporation of free RNA nucleotides into the growing chain (Fig. 1-17). However, not all of the DNA is copied into RNA, only those portions that encode genes. In eukaryotic cells, only a small fraction of the total DNA is actually used to encode genes. Furthermore, not all genes are transcribed into mRNA in equal amounts in all cells. The process of transcription is tightly regulated, so that only those mRNAs that encode the proteins that are currently needed by each cell are manufactured at any one time. This overall process is known as **gene expression.** Understanding the process of gene expression and how it differs in different types of cells and under different conditions is one of the fundamental questions driving the technologies of genomics.

The primary control of transcription takes place in a region of DNA known as the promoter, which occupies a position upstream (in the 5' direction) from the part of a gene that will be transcribed into RNA (the protein-coding region of the gene). There are a huge variety of different proteins that recognize specific DNA sequences in this promoter region and that bind to the DNA; they either assist or block the binding of RNA

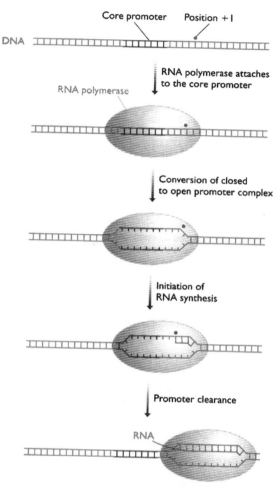

FIGURE 1-17. RNA polymerase II attaches to the promoter and begins transcription. Reproduced, with permission, from T. Brown, *Genomes* 2nd edn. Copyright 2002, BIOS Scientific Publishers Ltd.

polymerase (Fig. 1-18). These DNA-binding proteins work in a combinatorial fashion to provide fine-grained control of the expression of each gene, depending on the type of cell, where it is located in the body, its current metabolic condition, and responses to external signals from the environment or from other cells.

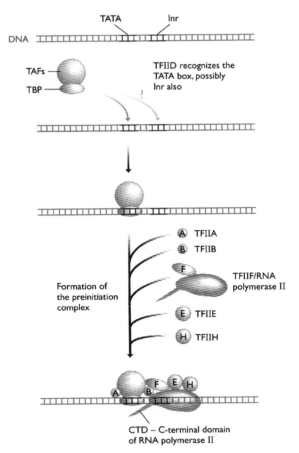

FIGURE 1-18. RNA polymerase II is actually a complex structure composed of many individual proteins. Reproduced, with permission, from T. Brown, *Genomes* 2nd edn. Copyright 2002, BIOS Scientific Publishers Ltd.

The factors governing the assembly of the set of proteins involved in regulating DNA transcription are much more complicated than just the sum of a set of DNA sequences neatly located in a promoter region, 5' to the coding sequence of a gene. In addition to the double helix, DNA has tertiary structures that involve twists and supercoils as well as winding around histone proteins. These 3-dimensional structures can bring distant regions of a DNA molecule into close proximity, so that proteins bound to these sites may interact with the proteins bound to the promoter region. These distant sites on the DNA that may effect transcription are known as enhancers. The total set of DNA-binding proteins that interact with promoters and enhancers are known as transcription factors, and the specific DNA sequences to which they bind are called transcription factor–binding sites.

RNA PROCESSING

Once a gene is transcribed into RNA, the RNA molecule undergoes a number of processing steps before it is translated into protein. First, a 5' cap is added, then a polyadenine tail is added at the 3' end. In addition, eukaryotic genes are broken up into protein coding **exon** regions separated by nonprotein coding **introns,** which are spliced out. This splicing is sequence specific and highly precise, so that the final product contains the exact mRNA sequence that codes for a specific protein with not a single base added or lost (Fig. 1-19).

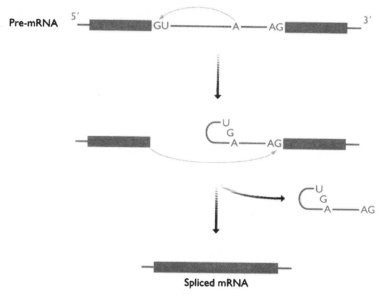

FIGURE 1-19. Intron splicing forms a mature mRNA from a pre-mRNA transcript. Reproduced, with permission, from T. Brown, *Genomes* 2nd edn. Copyright 2002, BIOS Scientific Publishers Ltd.

Each of these posttranscriptional processes may serve as a point of regulation for gene expression. Capping, polyadenylation, and/or splicing may be blocked or incorrect splicing may be promoted under specific metabolic or developmental conditions. In addition, splicing may be altered to produce different mRNA molecules.

ALTERNATIVE SPLICING

Each gene does not encode a single protein, as was originally suggested by the studies of *Neurospora* enzymes by Beadle and Tatum. In many cases, there are several alternate forms of final spliced mRNA that can be produced from a single pre-mRNA transcript, potentially leading to proteins with different biological activities. In fact, current estimates suggest that every gene has multiple alternate splice forms. Alternate splicing my involve the failure to recognize a splice site, causing an intron to be left in, or an exon to be left out. Alternate splice sites may occur anywhere, either inside exons or introns, so that the alternate forms of the final mRNAs may be longer or shorter, contain more or fewer exons, or include portions of exons (Fig. 1-20). Thus each different splice form produced from a gene is a unique type of mRNA, which has the potential to produce a protein with different biochemical properties.

It not clear how alternative splicing is controlled. It may be that the signals that govern RNA splicing are not perfectly effective, or RNA splicing may be actively used as a form of gene regulation. It is entirely possible for the products of other genes to interact with RNA splicing factors—perhaps in conjunction with external signals—to alter RNA splicing patterns for specific genes. The net result is many different forms of mRNA, some produced only under specific circumstances of development, tissue specificity, or environmental stimuli. Thus, under some conditions, a different protein with an added (or

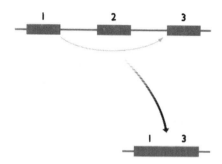

(A) Exon skipping

(B) Cryptic splice site selection

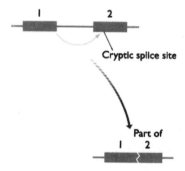

FIGURE 1-20. Two forms of splicing. Reproduced, with permission, from T. Brown, *Genomes* 2nd edn. Copyright 2002, BIOS Scientific Publishers Ltd.

removed) functional domain is produced from a gene, resulting in different protein function.

"Alternative splicing increases protein diversity by allowing multiple, sometimes functionally distinct proteins to be encoded by the same gene." (Sorek, 2001). The totality of all of these different mRNAs is called the *transcriptome,* which is certainly many times more complex than the genome. The relative levels of alternate splice forms for a single gene may have substantial medical significance. There are 60 kinase enzymes that have alternate splice forms that do not include their catalytic domains, creating proteins that may function as competitive inhibitors of the full-length proteins (Sorek, 2001).

TRANSLATION

For a gene to be expressed, the mRNA must be translated into protein. The theory behind this process was encapsulated quite neatly in 1957 by Crick's diagram of the Central Dogma, but the details of the information flow from DNA to mRNA to protein took another decade to work out. It was immediately clear that the cell must solve several different problems of information storage and transmission. Huge amounts of information must be stored in the simple 4-letter code of DNA, it must be translated into the quite different 20-letter code of amino acids, and a great deal of punctuation and regulatory information must also be accounted for.

The problem of encoding 20 different amino acids in the 4-letter DNA/RNA alphabet intrigued information scientists, physicists as well as biologists, and many ingenious incorrect answers were proposed. The actual solution to this problem was worked out with brute force biochemistry by Har Gobind Khorana (Soll et al., 1965) and Marshall W. Nirenberg (Nirenberg et al., 1965) by creating an **in vitro** (test tube) system in which pure pieces of RNA would be translated into protein. They fed the system with RNA molecules of simple sequence and analyze the proteins produced. With several years of effort (1961–1965), they defined a code of 64 three-base RNA **codons** that corresponded to the 20 amino acids (with redundant codons for most of the amino acids) and three "stop" codons that signal the end of protein synthesis (Table 1-1). Also in 1965, Robert W. Holley established the exact chemical structure of tRNA (transfer RNA), the adapter molecules that carry each amino acid to its corresponding codon on the mRNA. There is one specific type of

TABLE 1-1. TRANSLATION TABLE FOR THE EUKARYOTIC NUCLEAR
GENETIC CODE.

	U	C	A	G
U	UUU = phe UUC = phe UUA = leu UUG = leu	UC* = ser	UAU = tyr AUC = tyr UAA = stop UAG = stop	UGU = cys UGC = cys UGA = stop UGG = trp
C	CU* = leu	CC* = pro	CAU = his CAC = his CAA = gln CAG = gln	CG* = arg
A	AUU = ile AUC = ile AUA = ile AUG = met (start)	AC* = thr	AAU = asn AAC = asn AAA = lys AAG = lys	AGU = ser AGC = ser AGA = arg AGG = arg
G	GU* = val	GC* = ala	GAU = asp GAC = asp GAA = glu GAG = glu	GG* = gly

tRNA that binds each type of amino acid, but each tRNA has an
anticodon that can bond to several different mRNA codons.
(Holley, Khorana, and Nirenberg shared the 1968 Nobel Prize in
physiology or medicine for this work.)

The translation process is catalyzed by a complex molecular
machine called a ribosome. The ribosome is composed of both
protein and rRNA (ribosomal RNA) elements. Proteins are
assembled from free amino acids in the cytoplasm, which are
carried to the site of protein synthesis on the ribosome by
the tRNAs. The tRNAs contain an anticodon region that matches
the three nucleotide codons on the mRNA. Each tRNA attaches
to the anticodon, the amino acid that it carries forms a bond with

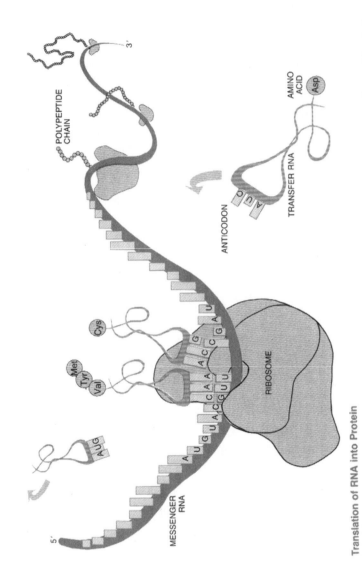

Translation of RNA into Protein

FIGURE 1-21. The ribosome interacting with tRNAs as it translates an mRNA into a polypeptide chain.

the growing polypeptide chain, the tRNA is released, and the ribosome moves down the mRNA to the next codon. When the ribosome reaches a stop codon, the chain of amino acids is released as a complete polypeptide (Fig. 1-21).

REFERENCES

Avery OT, MacLeod CM, McCarty M. Studies on the chemical nature of the substance inducing transformation of pneumococcal types. J Exp Med 1944;79:137–158.

Beadle GW, Tatum EL. Genetic control of biochemical reactions in Neurospora. Proc Natl Acad Sci USA 1941;27:499–506.

Chargaff E. Chemical specificity of nucleic acids and mechanisms of their enzymatic degradation. Experientia 1950;6:201–209.

Crick FHC. Nucleic acids. Sci Am 1957;197:188–200.

Hershey AD, Chase M. Independent functions of viral proteins and nucleic acid in growth of bacteriophage. J Gen Physiology 1952;36:39–56.

Holley RW. Structure of an alanine transfer ribonucleic acid. JAMA 1965;194:868–871.

Johannsen W. Elementeder exakten Erblichkeitslehre. Fischer, Jena, 1909; 516 pp. Summarized in: Strickberger HW. Genetics, Second Edition, MacMillan 1976; p. 276.

Lander ES, Linton LM, Birren B, et al. Initial sequencing and analysis of the human genome. Nature 2001;409:860–921.

Mendel, G. Versuche über Pflanzen-Hybriden. Verhandlungen des naturforschenden Vereines, Abhandlungen, Brünn 1866;4:3–47.

Meselson M, Stahl FW. The replication of DNA in *Escherichia coli*. Proc Natl Acad Sci USA 1958;44:671–682.

Morgan TH. The physical basis of heredity. Philadelphia: Lippincott, 1919.

Morgan TH. Sex-limited inheritance in *Drosophila*. Science 1910;32:120–122.

Nirenberg M, Leder P, Bernfield M, et al. RNA code words and protein synthesis, VII. On the general nature of the RNA code. Proc Natl Acad Sci USA 1965;53:1161–1168.

Mural RJ, Adams MD, Meyers EW, et al. The sequence of the human genome. Science 2001;291:1304–1351.

Pauling L, Corey R. Atomic coordinates and structure factors for two helical configurations of polypeptide chains. Proc Natl Acad Sci USA 1951;37:235–240.

Sayre A. Rosalind Franklin and DNA. New York: Norton, 1975.

Soll D, Ohtsuka E, Iones DS, et al. Studies on polynucleotides, XLIX. Stimulation of the binding of aminoacyl-sRNAs to ribosomes by ribotrinucleotides and a survey of codon assignments for 20 amino acids. Proc Natl Acad Sci USA 1965;54:1378–1385.

Sorek R, Amitai M. Piecing together the significance of splicing. Nat Biotechnol 2001;19:196–???.

Sutton W. The chromosomes in heredity. Biol Bull 1903;4:231–251.

Watson JD, Crick FHC. A structure for deoxyribose nucleic acid. Nature 1953;171:737.

GENOMIC TECHNOLOGY

CUT, COPY, AND PASTE

Genomics technology is all about the application of automation, and massively parallel systems to molecular biology. However, to understand these new high-throughput technologies, it is first necessary to understand the basic molecular biology techniques on which they are based. If we extend the metaphor of the genomic DNA sequence as a book, introduced in the previous chapter, then molecular biology provides the "cut," "paste," and "copy" operations needed to edit the text. Traditional molecular biology works on one gene ("word") at a time, whereas genomics technologies allow operations on all of the genes at once ("global search and replace").

RESTRICTION ENZYMES

It turns out that cutting DNA at specific positions is quite simple. Bacteria do it all the time using proteins called type II restriction endonuclease enzymes, which were first purified and characterized by Hamilton Smith in the early 1970's (Smith and Wilcox, 1970). It is possible to grow almost any strain of

Essentials of Medical Genomics, Edited by Stuart M. Brown.
ISBN 0-471-21003-X.   Copyright © 2003 by Wiley-Liss, Inc.

bacteria in a flask of nutrient broth, collect the bacterial cells, grind them up, and extract active restriction enzymes from the resulting goop of cellular proteins. A number of companies now specialize in purifying these enzymes and selling them to molecular biologists (at surprisingly inexpensive prices, considering their remarkable powers).

Each strain of bacteria makes its own characteristic restriction enzymes, which cut DNA at different specific sequences,

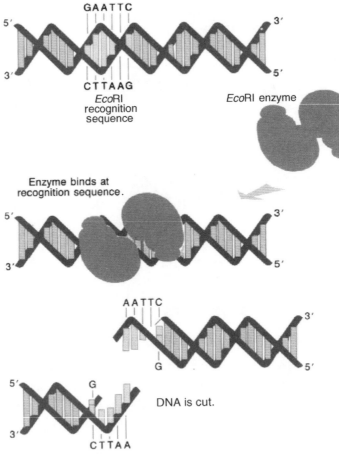

FIGURE 2-1. The *Eco*RI restriction enzyme produces sticky ends. (Art concept developed by L. Shoemaker.)

known as recognition sites, which are typically four, six, or eight bases long. Not surprisingly, bacteria protect themselves from their own restriction enzymes by avoiding the use of the recognition sequence in their own DNA and/or by the action of sequence specific DNA methylase enzymes, which modify the DNA at the recognition site so that it cannot be cut.

In addition to having a specific DNA sequence that it recognizes as its cleavage site, each type of restriction enzyme cuts the DNA in a specific pattern, leaving a characteristic shape at the free ends. Most restriction enzymes cut the two strands of the DNA double helix unevenly, leaving a few bases of over-hang on one strand or the other (Fig. 2-1). The overhanging bases

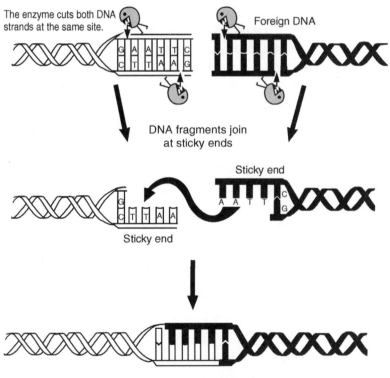

FIGURE 2-2. Two different pieces of DNA are cut with *Eco*RI. The fragments are ligated to create a recombinant molecule.

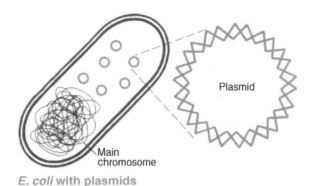

E. coli with plasmids

FIGURE 2-3. A plasmid is a circle of DNA maintained independently from the chromosome within a bacterial cell.

from the two freshly cut ends are complementary in sequence, so under the right conditions, they can pair back up to form new hydrogen bonds. These are known as sticky ends.

Another bacterial enzyme, known as DNA ligase, can re-create the phosphate bonds of the DNA backbone across a pair of rejoined sticky ends, effectively pasting together a new DNA molecule. With enzymes that can cut and paste, it is not tremendous leap to the concept of cutting two different pieces of DNA with the same restriction enzyme, then swapping fragments and splicing them with ligase in a new combination (Fig. 2-2). Paul Berg made the first artificial recombination in 1972 between a piece of SV40 virus and a piece of *Escherichia coli* chromosomal DNA (Jackson et al., 1972).

DNA CLONING IS COPYING

Cloning is a process of making identical copies by biological duplication. DNA cloning makes use of bacteria as the hosts to grow unlimited copies of a single piece of DNA. Bacteria have a single circular chromosome, but they also have some additional small circles of DNA called **plasmids** (Lederberg, 1952; Fig. 2-3), which carry genes for functions that evolve rapidly, such as antibiotic resistance (Ochiai et al., 1959). Each bacterial cell has just one copy of its chromosome, but it may have hundreds of copies of a plasmid. Bacteria have natural mechanisms for transferring plasmids from one cell to another and taking up plasmids from the environment.

Just as Berg was experimenting with cutting and splicing DNA using restriction enzymes, Stanly Cohen developed a techique to insert plasmids carrying antibiotic resistance genes into bacteria. Cohen and Herbert Boyer used plasmids as **cloning vectors** to carry fragments of DNA from other organisms into bacteria where they could make millions of copies of those fragments (Cohen et al., 1973). The cloning process begins by isolating DNA from an organism of interest and cutting it using a **restriction enzyme,** which cuts the DNA at a specific sequence, such as GAATTC. Plasmid DNA is isolated from bacteria, and it is also cut with the same restriction enzyme, opening the DNA circle. Then the cut fragment of interesting DNA is mixed with the cut plasmid DNA, and they are joined together, using DNA **ligase,** into a new circular molecule that contains both plasmid and the target DNA. This new molecule is called a **chimeric** plasmid because it is made up of DNA from two different types of organisms. The first chimeric clone was created in 1973 by splicing DNA from a frog into an *E. coli* plasmid by Cohen, Chang, Boyer and Helling. They further demonstrated that *E. coli* will produce foreign proteins from genes cloned into plasmids (Morrow et al., 1974).

The chimeric plasmid is then put back into a bacterial cell using a process called **transformation,** basically using the bacterium's natural ability to take up plasmid DNA from a solution. The bacterial cells carrying the chimeric plasmid are then put into a medium where they can grow; and as the bacteria multiply, so does the plasmid (Fig. 2-4). Then the bacteria are harvested, and large quantities of plasmid with cloned DNA can be purified. The interesting DNA fragment can be removed from the plasmid DNA by again cutting with a restriction

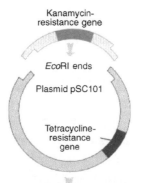

Kanamycin-
resistance gene

*Eco*RI ends

Plasmid pSC101

Tetracycline-
resistance
gene

Fragments joined with
DNA ligase.

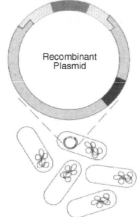

Recombinant
Plasmid

E. coli cell transformed
with recombinant plasmid.

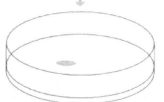

Transformed cells plated
onto medium containing
tetracycline and kanamycin.

Only cells containing recombinant
plasmid survive to produce
resistant colony.

FIGURE 2-4. Ligation of a foreign gene into a plasmid and cloning in *E. coli*
cells.

enzyme. Some bacteria containing the chimeric plasmid can also be frozen so that more cloned DNA can be obtained whenever it is needed.

These methods for cutting, pasting, and copying DNA can be used to construct complex DNA molecules that have parts from several different genes or from different organisms. Taken together, this technology is called **recombinant DNA cloning** or **genetic engineering.** Some viruses that grow in bacteria (bacteriophage) can also be used as cloning vectors by replacing part of the natural virus DNA with some other piece of DNA. To work with large fragments of DNA, vectors have been developed that act as artificial chromosomes in bacteria (bacterial artificial chromosomes; BACs) or yeast (yeast artificial chromosomes; YACs).

PCR IS CLONING WITHOUT THE BACTERIA

Molecular biology is not a discipline for people who expect to spend a few years learning a set of skills and then to sit back and use them for a few decades. Just when you think that you know all the basic chops in the laboratory, somebody comes along and reinvents the entire field. Polymerase chain reaction (PCR) was such a technical revolution. The basic concept is simple: Use the DNA polymerase that organisms use to copy their own DNA to copy specific pieces of DNA. Target the copying process by using a short primer that is complementary to one end of the desired sequence; then make copies of the other strand by using a second primer that is complementary to the other end of the desired sequence. Then make another copy in the forward direction, another copy in the reverse direction, and repeat for many cycles. After the first cycle, the newly synthesized fragments serve as templates for additional rounds of copying. The net result is that each round of copying doubles the number of copies of the desired DNA fragment, leading to an exponential

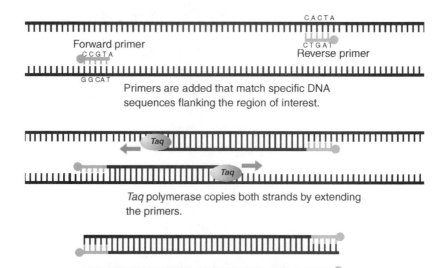

CACTA

Forward primer
CCGTA
GGCAT

CTGAT
Reverse primer

Primers are added that match specific DNA
sequences flanking the region of interest.

Taq polymerase copies both strands by extending
the primers.

After a number of cycles, many copies of the fragment are created.

FIGURE 2-5. In PCR, a pair of primers complementary to the two ends of a target DNA sequence bind to the DNA, and complementary copies are synthesized by *Taq* DNA polymerase. Each new strand serves as a template for additional rounds of synthesis, allowing the creation of large amounts of the target fragment.

amplification (Fig. 2-5). Even if the amplification is not perfectly efficient, millions of copies are created in about 20 cycles.

The ingredients for the PCR reaction were available in the typical molecular biology laboratory for 10–15 years before Kary Mullis worked out the technique in 1983 (Saiki et al., 1988). A lot of scientists were slapping themselves on the forehead when Mullis picked up the Nobel Prize in 1993. The one additional element that made PCR simple and user-friendly was the discovery that bacteria, such as *Thermophilus acquaticus,* that live in hot springs and deep ocean thermal jets, have heat-resistant DNA polymerase enzymes (e.g., *Taq* polymerase). These heat-stable enzymes allow the PCR reaction to proceed for many cycles by simply heating and cooling a tube with the target DNA, the two primers, polymerase, and the free G, A, T, and C nucleotide triphosphates.

The beauty of the PCR process is not just that it makes lots of copies of a DNA fragment in a simple single-tube reaction. It can also be used to pull out a single specific DNA fragment from a complex mixture, such as an entire genome. PCR can be used to amplify substantial amounts of specific DNA fragments, which can be used for other molecular techniques, from tiny and impure samples. such as found in clinical diagnostics for infectious agents, in forensic investigations, and in fossil remains. The basic requirement is that about 20 bases of sequence must be known at each end of the fragment of DNA that is to be amplified so the forward and reverse primers can be created. However, molecular biologists have worked out dozens of clever methods that allow PCR amplification when only a single

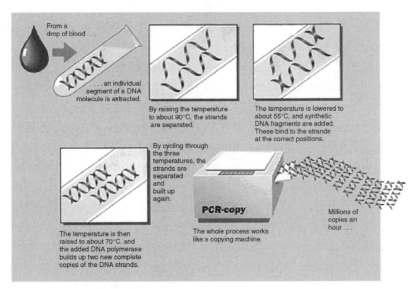

FIGURE 2-6. The PCR process used for a diagnostic test. (Modified from the Nobel Foundation Web site; http://www.nobel.se/chemistry/laureates/1993/illpres/pcr.html).

primer sequence is known, when the sequences flanking the desired DNA segment are only partially known, and/or when new flanking sequences are attached to an unknown target in some complex cloning scheme.

PCR is an essential ingredient in many different DNA diagnostic tests. It allows a small sample of patient DNA (or RNA) to be used as source material to generate sufficient quantities of specific DNA fragments that can be sequenced; identified by mass spectroscopy; or detected by a variety of other labeling, probe, and visualization schemes (Fig. 2-6).

GENOME SEQUENCING

It is a bit of a conceptual leap from the discovery of the structure of DNA to the sequencing of the human genome, but one leads directly to the other. The double-helix model of DNA led to an understanding of how the DNA is duplicated as cells grow and divide. This process of DNA replication was then harnessed as a tool for the Sanger method of determining the sequence of a piece of DNA (see below).

Modern DNA sequencing technology is based on the method of controlled interruption of DNA replication developed by Fred Sanger in 1977 (Sanger et al., 1977) (for which he was awarded the Nobel Prize in 1980 together with Walter Gilbert and Paul Berg). Sanger combined the natural DNA replication machinery of bacterial cells with a bit of recombinant DNA technology and some clever biochemistry to create an *in vitro* system in which a cloned fragment of DNA is copied, but some of the copies are halted at each base pair along the sequence. Natural DNA replication uses a DNA polymerase enzyme that copies a template DNA sequence (one half of the DNA double helix) and creates a new DNA polymer, complementary to the template, by joining free deoxynucleotides into a growing DNA chain. The replication reaction also requires a primer—a short piece of DNA that is complementary to the template—to which the polymerase can affix the first added base (Fig. 2-7).

THE SANGER METHOD

The Sanger sequencing method makes use of specially modified dideoxynucleotides which stop (terminate) the replication process if they are incorporated in the growing DNA chain instead of the normal deoxynucleotide. For each template (piece of cloned DNA to be sequenced), four separate sequencing reactions are set up that each contain one of the dideoxynucleotides (ddG, ddA, ddT, and ddC) as well as a full set of normal deoxynucleotides, the primer, and the DNA polymerase enzyme. For example, in the reaction mixture containing ddA, some of

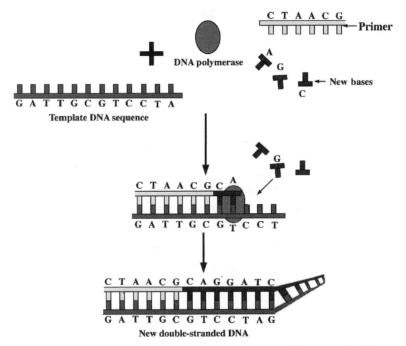

FIGURE 2-7. DNA polymerase uses a primer and free nucleotides to synthesize a complementary strand for a template DNA sequence.

the growing strands are stopped when they reach each A in the template sequence. The resulting set of DNA fragments form a nested set, all starting at the same point, but ending at different A residues. Similar reactions are set up to stop replication at G, T, and C residues.

To visualize the resulting DNA fragments, it is necessary to incorporate some type of labeled molecule, usually radioactive, in the replication reaction. It is possible to label the primer, the deoxynucleotides, or the dideoxynucleotides. In any case, the fragments are separated by length using polyacrylamide gel electrophoresis, with one gel lane for each of the four different dideoxynucleotide reactions. DNA fragments of a specific length form a distinct band on the gel, so there is one band for each base in the template sequence. Then the gel, which contains the radioactively labeled DNA fragments, is placed on top of a sheet of X-ray film so that the radioactive bands of DNA can expose it (Fig. 2-8). Finally, the sequence is manually read off of the X-ray film from the positions of the bands and typed into a computer (Fig. 2-9).

The value of determining DNA sequences was immediately obvious to many biologists, but the laboratory techniques of the Sanger method are both laborious and technically demanding. DNA sequencing became a rite of passage for many molecular biology graduate students in the 1980s and early 1990s. Initially, some kits were developed to simplify and standardize the biochemistry. These kits eventually included superior types of polymerase enzymes, and minor improvements were made in the polyacrylamide gel apparatus; however, the essential technique remained unchanged for about 15 years.

Automated DNA Sequencing

The first major innovation to improve DNA sequencing was Leroy Hood's development of fluorescently labeled nucleotides

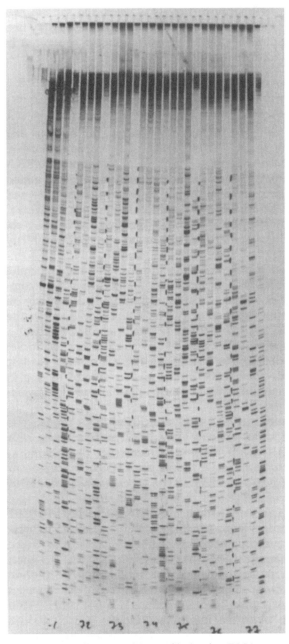

FIGURE 2-8. An autoradiogram (X-ray film) of a DNA sequencing gel. Each sequence requires four lanes, one for each base.

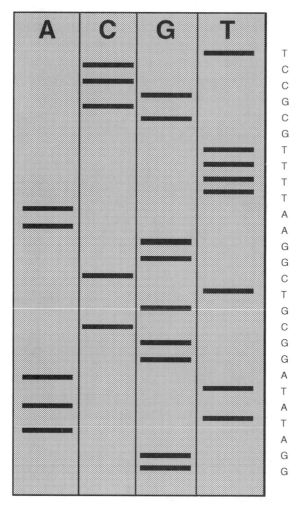

FIGURE 2-9. A sequencing gel showing bands in four gel lanes, representing the DNA fragments produced by the four different dideoxy sequencing reactions. The final DNA sequence is shown at the right.

in 1985 to replace the standard radioactive labels (Smith et al., 1986). The fluorescent labels could be measured directly in the acrylamide gel as DNA fragments passed by a laser/detector, thus eliminating both the radioactivity and the X-ray film. In addition, Hood used four different colored fluorescent labels, one for each of the four DNA bases, so that after the sequencing reactions were completed, the four sets of fragments could be run in a single lane of an acrylamide gel and the base determined by the color of each fragment (Fig. 2-10; color insert).

Hood also directly connected the fluorescent detector to a computer so that the fluorescent signal was automatically collected and converted to a DNA sequence (Fig. 2-11). Together with Lloyd Smith, Michael Hunkapiller, and Tim Hunkapiller, Hood founded Applied Biosystems, Inc. (ABI), which manufactures a commercial version of this fluorescent sequencer, which became available in 1986. Since 1986, ABI (in cooperation with the PerkinElmer Corp.) has consistently improved their machines and has dominated the commercial marketplace for automated sequencers. Essentially all of the Human Genome Project and absolutely all of Celera Genomics' sequencing was done on ABI machines. However, ABI machines still have many of the limitations of the original Sanger method. They still rely on DNA polymerase to copy a template DNA sequence and on polyacrylamide gel electrophoresis to separate the fragments.

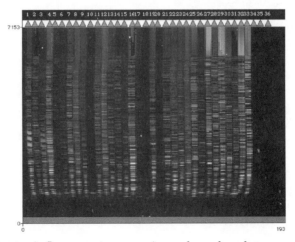

FIGURE 2-10. A fluorescent sequencing gel produced on an automated sequencer. Each lane contains all four bases, differentiated by color. Figure also appears in Color Figure Section.

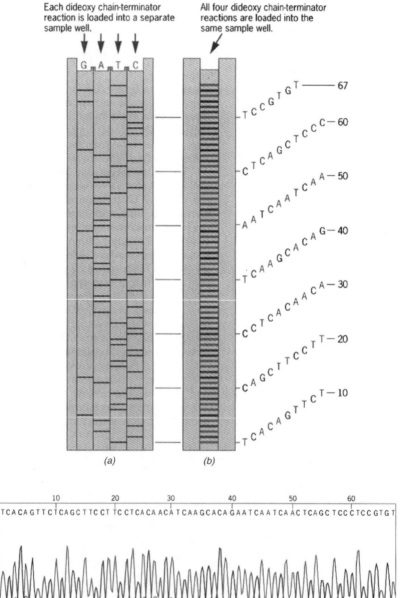

FIGURE 2-11. ABI fluorescent sequencers allow all four bases to be sequenced in a single gel lane and include automated data collection. Figure also appears in Color Figure Section.

SUBCLONING

One of the key limitations of the Sanger/ABI method is that DNA sequences can be determined only in chunks of 500–800 bases at a time (known as reads). Larger fragments cannot be resolved by polyacrylamide gel electrophoresis. As a result, determining the sequence of large pieces of DNA requires sequencing many overlapping fragments and then assembling them. There are a variety of strategies for breaking up large DNA molecules—such as human chromosomes—into overlapping fragments for sequencing.

Early DNA sequencing projects proceeded methodically. A scientist would first map a chunk of DNA for various restriction enzyme cut sites and then clone these restriction fragments into plasmids (subcloning), carefully tracking how they would reassemble. This process needed to be done at least twice to generate two sets of overlapping fragments, since the sequences at the ends of reads tend to have many errors. Alternately, sets of nested fragments could be generated by chewing away at one end of a cloned fragment with a DNase enzyme and stopping the reaction at various time points. Then the chewed fragments would be subcloned; next their sizes would be determined, and the fragments sequenced.

All of this laboratory cloning work was time-consuming and could not be scaled up for larger projects. As the cost of automated sequencing went down and the speed and through-put of the machines increased, it became necessary to find faster methods of generating small fragments to be sequenced. In theory, if a chunk of DNA is copied many times (cloned) and all of the copies are broken up into many random fragments (shotgun subcloning) and the fragments are sequenced, then eventually a complete set of overlapping fragments can be found and the sequences of these fragments can be assembled into the original chunk. This ends up being something of a statistical game. To find a set of overlapping fragments that completely cover a large chunk of DNA by sequencing random shotgun subclones, it is necessary to sequence a total amount of DNA that is much more than the overall length of the original DNA chunk. The shotgun clones form a Poisson distribution—sort of like trying to hit all of the squares of a chessboard by hanging it on the wall and throwing darts: You will hit some squares many times before all of them are hit.

In the mid-1990s, the Institute for Genomic Research (TIGR) became a champion of the shotgun style of sequencing. It streamlined the process of randomly cloning thousands of fragments from large DNA molecules and efficiently feeding clones into a room full of automated sequencers, minimizing the number of scientist-hours spent generating the sequence reads. Then it devoted the bulk of its efforts into developing good sequence assembly software and finishing the assembly of sequences using a group of computer-savvy scientists to help the software, and occasionally going back into the lab to resequence troublesome spots. The success of this method was dramatically demonstrated by TIGR's publication of the complete sequence of the *Haemophilus influenzae* genome at a time when the rest of the scientific community was sure that such a large project could not be done by shotgun sequencing and was, in fact, beyond the reach of current technology (Fleischmann et al., 1995).

SEQUENCE ASSEMBLY

The assembly of shotgun fragments is obviously a job for computer software, but there are some problems associated with the data from automated sequencers. DNA sequencers are not perfectly accurate, and mistakes are much more common at the beginning and the ends of each sequence read—precisely the regions where overlaps are most likely to be found. DNA contains many types of repeats, ranging from long tracts of a single base or a simple repeat of two or three bases to tandem (or inverse) repeats hundreds to thousands of bases long. Repeats make it difficult to assemble overlapping sequences unless a single read spans an entire repeat and includes the nonrepetitive sequence on both sides.

There are some additional problems with the assembly of huge genome sequences, particularly eukaryotes, that did not affect scientists working on the sequencing of smaller pieces of DNA. Eukaryotic genomes have duplicated genes, and even duplications of entire sections of chromosomes. Eukaryotic chromosomes contain centromeres and telomeres, which consist of nothing but thousands of repeats of the same short sequences. The human genome also contains about 100,000 copies of a sequence called ALU, which is an inactive transposon 147 bases long. Clearly, a sequence fragment that ends in the middle of an ALU sequence could overlap and assemble with any other ALU-containing fragment, possibly from another chromosome, leading to incorrect assemblies.

SEQUENCING THE HUMAN GENOME

The public Human Genome Project (HGP) consortium has relied primarily on a map-based strategy to sequence the human genome (Lander et al., 2001). It spent several years cloning 100,000 to 1,000,000 base-pair chunks of human DNA into large vectors called BACs; then the scientists painstakingly assembled a complete set of overlapping BACs that covered every chromosome (known as a tiling path). Only after this BAC map was completed, did they start large-scale sequencing of each BAC by breaking it into many small fragments that could be directly sequenced by the Sanger method on automated DNA sequencers. Finally, all of the short sequences had to be

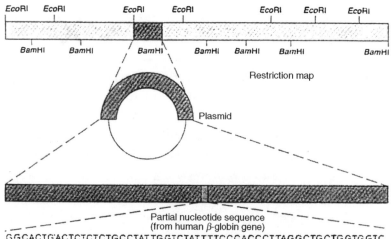

FIGURE 2-12. Subcloning of restriction fragments from a chromosome into a plasmid and then sequencing the fragments in chunks of 500–600 bases.

assembled back into complete chromosomes, using computer programs (Fig. 2-12).

In contrast, the Celera Genomics Corporation developed a human genome sequencing strategy relied entirely on a shotgun sequencing approach (Venter et al., 2001). Rather than carefully building a set of large overlapping BAC clones that covered the entire genome, they randomly cut genomic DNA, taken from five different people, into fragments, and sequenced a number of fragments equivalent to six times the total size of the human genome (i.e., 18 billion bases of sequence information). Celera then used custom designed computer programs to assemble the overlapping fragments.

Both the HGP and the Celera approach lead to an uneven sequence of the genome. Some areas are sequenced many times over, but there are still gaps. There is also a point of diminishing returns at which more random sequencing is not likely to fill stubborn gaps that might be caused by special properties of the DNA in those particular locations. For example, centromeres and telomeres contain highly repetitive sequences that resist cloning and cannot by assembled by computer algorithms. At that point, which has already been reached for the human genome sequence, the only way to fill the gaps is by changing the strategy to a much more painstaking, hands-on approach. Small teams of biologists must tackle the unique problems of each gap. It will be a few more years before the human genome sequence is perfectly complete without any gaps.

REFERENCES

Cohen SN, Chang AC, Boyer HW, Helling RB. Construction of biologically functional bacterial plasmids in vitro. Proc Natl Acad Sci USA 1973;70:3240–3244.

Fleischmann RD, Adams MD, White O, et al. Whole-genome random sequencing and assembly of *Haemophilus influenzae Rd*. Science 1995;269:496–512.

Jackson DA, Symons RH, Berg P. Biochemical method for inserting new genetic information into DNA of simian virus 40: Circular SV40 DNA molecules containing lambda phage genes and the galactose operon of *Escherichia coli*. Proc Natl Acad Sci USA 1972;69:2904–2909.

Lander ES, Linton LM, Birren B, et al. Initial sequencing and analysis of the human genome. Nature 2001;409:860–921.

Lederberg J. Cell genetics and hereditary symbiosis. Physiol. Rev. 1952;32: 403–430.

Morrow JF, Cohen SN, Chang AC, et al. Replication and transcription of eukaryotic DNA in *Escherichia coli*. Proc Natl Acad Sci USA. 1974;71:1743–1747.

Ochiai K, Yamanaka T, Kimura K, Sawada O. Studies on the inheritance of drug resistance between *Shigella* strains and *Escherichia coli* strains. Nilon Iji Shimpo 1959;1861:34–46.

Saiki RK, Gelfand DH, Stoffel S, et al. Primer-directed enzymatic amplification of DNA with a thermostable DNA polymerase. Science 1988;239:487–491.

Sanger F, Nicklen S, Coulson AR. DNA sequencing with chain-terminating inhibitors. Proc Natl Acad Sci USA 1977;74:5463–5467.

Smith HO and Wilcox KW. A restriction enzyme from *Haemophilus influenzae*. I. Purification and general properties. J Mol Biol 1970; 51:379–391.

Smith LM, Sanders JZ, Kaiser RJ, et al. Fluorescence detection in automated DNA sequence analysis. Nature 1986;321:674–679.

Venter JC, Adams MD, Myers EW, et al. The sequence of the human genome. Science 2001;291:1304–1351.

BIOINFORMATICS TOOLS

PATTERNS AND TOOLS

The success of the Human Genome Project (HGP) and the foundation of the entire field of genomics is based on the automation of biochemical laboratory methods. This is the same basic concept used for robotic welders in automobile factories. However, the product of automated high-throughput genomics laboratories is information rather than cars. Vast quantities of information. This information requires specialized tools for storage and analysis—and that is what bioinformatics is all about.

Bioinformatics is the use of computers for the acquisition, management, and analysis of biological information. It is a hybrid discipline that requires skills from molecular biology, computer science, statistics, and mathematics. In practice, bioinformatics specialists may also need to know a fair bit about computer hardware, networking, robotics, image processing, and anything else that affects the collection, storage, analysis, and distribution of biological information.

The average biologist has been forced to learn a lot about bioinformatics since the early 1990s. The use of DNA and protein sequence data has become part of the routine daily

Essentials of Medical Genomics, Edited by Stuart M. Brown.
ISBN 0-471-21003-X.   Copyright © 2003 by Wiley-Liss, Inc.

work in most biology laboratories—an essential component of both experimental design and the analysis of results. It seems unlikely that use of computer tools for the manipulation of sequence data is destined become part of the routine work of the typical medical doctor. However, just as doctors must know the procedure to test a throat culture for *Strep.*, even when the actual microbiology labwork is done at a commercial lab, so will they need to understand how DNA sequences are used for diagnostic and therapeutic purposes—both the theory and the practical aspects of the technology. With that in mind, this chapter describes a variety of bioinformatics tools in sufficient detail to allow a solid understanding of how they work, but it is not a detailed tutorial on their use.

The current set of commonly used bioinformatics tools was not derived from some coherent set of fundamental theoretical principles. On the contrary, the bioinformatics toolkit is a hodgepodge collection of unrelated algorithms that have been borrowed from many different branches of mathematics, linguistics, computer science, and other disciplines and then modified through multiple generations of trial-and-error improvement. The current set of bioinformatics tools are what seem to work best to solve a number of different practical problems; but, at any time, a new tool may emerge from some totally unexpected theoretical background that works better for some specific task. Furthermore, bioinformaticians are continually scrambling to find tools to deal with new types of biological information, such as the products of new genomics technolgies. Bioinformaticists are not picky, they will use whatever tool works best for a specific job.

In no particular order, these tools include programs to draw maps of plasmids containing cloned genes, search DNA sequences for sites (short patterns) that can be cut by specific restriction enzymes, design polymerase chain reaction (PCR) primers that can be used to target a specific fragment of DNA, compare one

DNA or protein sequence to another or search an entire database of sequences for similarity to one query sequence, line up two sequence or a group of sequences (**multiple alignment**), join overlapping fragments of DNA sequences (sequence assembly), predict the chemical and structural properties of a protein from its amino acid sequence, predict the function of a new protein based on its containing subsequences conserved in known protein families (**motifs**), and calculate a tree of evolutionary relationships among a set of sequences (**phylogenetics**).

Listed out like this, there seems to be no commonality to these tools whatsoever. That is not true; there are a few consistent ideas common to most of the tools. The most basic is the idea of pattern recognition (Fig. 3-1). The pattern may be as simple as the four to six bases that define a restriction enzyme recognition site, or as complex as a conserved structural domain in G-protein-coupled receptors. The problem of computer pattern recognition has been tackled independently by many different disciplines, ranging from military remote sensing to voice recognition for collect calls from a pay phone. Many of these different approaches have been tried out for various bioinformatics problems. The current crop of tools used in bioinformatics have gone through some evolution and selection, but they are by no means optimal solutions to these many diverse problems. In many cases, bioinformatics experts will repeat an analysis with several different tools because they don't trust that any one tool does the best job in all situations.

One thing that is not consistent is the amount of data that the various bioinformatics programs work on and the amount of

FIGURE 3-1. With computers, it's easy to find patterns, even if they are not really there. These letters can be found in butterfly's wings. (Adapted from *The Butterfly Alphabet*, Sandred KB. Scholastic 1999; 64pp.).

computing power that they require to execute their task. Some operations, such as finding restriction sites in a plasmid, can be handled in few seconds on any desktop computer, while making a similarity search of a DNA sequence against all of GenBank requires many gigabytes of hard disk space and gigaflops of computing power.

SEQUENCE COMPARISON

One of the most basic questions in bioinformatics is: How similar are these two sequences? Calculating sequence similarity is a deceptively difficult problem. For two short sequences, you might just write them down on two slips of paper, or type them in on two lines of a word-processing program, and then try to slide them by each other to see if there is any group of letters that lines up (Fig. 3-2).

The best overall similarity might require some mismatches or inserting some gaps in one or both sequences. Is a short identical region better than an overall match with a scattering of mismatches and gaps? You would think that some mathematicians must have worked on this problem and come up with an optimal way of calculating similarity; and indeed they have. In fact, calculating a similarity score for two sequences and finding the best alignment between them turns out to be one and the same problem, and a good solution to this problem has been available since the early 1970s (Needelman and Wunch, 1970).

Similarity scores can be calculated for both DNA and protein sequence pairs with one modification. DNA to DNA similarity is almost always scored either as an identical match or as a mismatch between two bases, with some penalty for inserting a gap

GATGCCAT**AGAGCGTAGTC**GTTCCCT   <—

—>   CTAG**AGAGCGTAGTC**AGAGTGTCTTTGAGTTCC

FIGURE 3-2. Aligning two sequences by hand.

in either sequence. However, when aligning two proteins, there can be various shades of similarity between a pair of amino acids that are not identical. Some pairs of amino acids have similar chemical properties whereas other pairs have similar codons— so they are separated by just a single mutation event (see Table 1-1). In practice, the best solution was found to be based on the frequency that one amino acid replaces another at an analogous position in sets of closely related proteins. This table of natural mutation rates between every possible pair of amino acids can then be used as a scoring matrix for amino acid sequence alignments (Dayhoff et al., 1978).

Needelman and Wunch described the use of the computational technique of dynamic programming for sequence alignment in a paper published in 1970, and this method was rapidly implemented in several simple computer programs. The Needelman–Wunch method finds the best overall alignment between two whole sequences (global alignment), but this does not always find the best alignment if just a small part in the middle of one sequence matches a part in the middle of another sequence. An improved method of **local alignment** using dynamic programming was developed by Smith and Waterman in 1981. A local alignment finds the best subsequence match between a pair of sequences (Fig. 3-3).

Local alignments are also used to answer another common bioinformatics question: What sequences in a database are most similar to this sequence? Or more generally: Is this sequence like anything that anyone else has ever seen before? This requires making a pairwise comparison between a query sequence and every sequence in the database, then choosing the database sequences that give the best overall alignment. The Smith-Waterman method finds the optimal match in a comparison between two sequences, but it is very slow when it is used to compare a sequence against all of the other sequences in a large database.

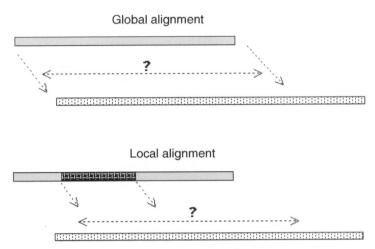

FIGURE 3-3. The difference between global and local alignment.

To make faster comparisons, some shortcuts had to be found. So far, the best methods for comparing a single sequence to a database involve quickly scanning each database sequence for short groups of letters (known as "words") that match the query sequence, then throwing away all of the sequences that do not have any good short matches. Then each short match is used as the start of a local alignment, which is extended in both directions until no more matches are found. A score is calculated for each alignment; then the database sequences that have the regions that align best are shown as the results of the search. This type of database search is called **heuristic** because it is approximate. The trade-off for greater speed is that it is possible to miss some important matches—particularly those with moderate overall similarity but no short regions of really high matching—and to get some (or many) false-positive matches.

Within this general area of heuristic searching for local alignments, a number of different computer programs have been developed. Pearson and Lipman (1988) created **FASTA,** a fast-alignment program, in the mid-1980s and have continued to refine it through many generations of optimization and improved functionality (Pearson, 1990). Altschul, Gish, Lipman, and others (1990) at the **National Center for Biotechnology Information** (NCBI) created a rival program called **BLAST** (Basic Local Alignment Search Tool; Altshul et al., 1990), which is even faster than FASTA, if sometimes a bit less sensitive.

Another important feature of both BLAST and FASTA is that they return an **E-value** (expected value), which is a statistical measure of the quality of each match. E-values are actually a measurement of how likely it would be to find a match of a similar quality if a search were done with a randomly generated sequence the same length as the query sequence. In more formal terms, an E-value is a measure of the probability that an observed match between two sequences is due to chance. In common language, an E-value is a measurement of the like-lihood that the match is bogus. However you look at it, a smaller number is a better match. E-values are generally small fractions expressed in exponential notation (i.e., 3.2×10^{-56}), so the larger the negative exponent, the smaller the E-value and the more significant the match. Typically, matches with E-values <0.05 are considered significant—just as P values are used in classic statistical tests. However, the E-value of a match depends on a number of factors including the length of the query sequence (short queries cannot give highly significant matches) and the size of the database being searched (the larger the database, the greater the chance for bogus matches).

BLAST has undergone many cycles of change and optimiza-tion (Altschul et al., 1997), and it is currently the most popular tool for comparing sequences. The NCBI operates a free BLAST server on its Web site (www.ncbi.nlm.nih.gov), which is used by many thousands of scientists each day. No one has suggested that either BLAST or FASTA provides an optimal, or even a good solution for similarity searching of databases; they are simply the best tools that are currently available. Some bioinfor-matics groups with lots of money have resurrected the

Smith-Waterman search method in custom-built supercomputers to make more precise and (they hope) more sensitive searches.

Using BLAST to compare pairs of sequences or to compare one sequence to a database is a common research task, but it is not typically used in diagnostic or forensic medicine. BLAST is generally used when a query sequence is completely unknown. Most medical applications of DNA sequence comparison involve looking for small changes in known sequences. If a direct comparison of sequences is made, it will usually be done by multiple alignment.

MULTIPLE ALIGNMENT

Sometimes it is necessary to make comparisons among a group of related sequences as a multiple alignment (Fig. 3-4). This is often important in the study of protein families and motifs. It is also the starting point for evolutionary studies (phylogenetics). In clinical medicine, it may also be done in the study of a particular gene in samples collected from many people. Since good algorithms and computer programs are available for aligning two sequences, one might expect that aligning groups of

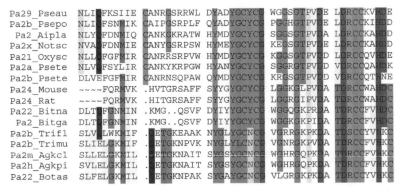

FIGURE 3-4. A multiple alignment of a part of the *Pa* gene from 15 different species.

sequences would just be an extension of these same methods. However, from a computational perspective, it is surprisingly difficult to align a group of sequences. The problem is that each additional sequence added to the alignment requires that adjustments be made to every other sequence (inserting new gaps, shifting mismatches to accommodate a new consensus, etc.). In fact, if a dynamic programming approach is used, each new sequence *exponentially* increases the amount of computing required to build the optimal alignment. The amount of computing gets huge if more than about 10 sequences are aligned.

Instead of using a dynamic programming approach to calculate the absolutely optimal alignment of a group of sequences, a shortcut approach called **progressive pairwise alignment** has been developed (Feng and Doolittle, 1987). This method relies on a quick set of pairwise comparisons among all of the sequences to be aligned to estimate the relative amounts of similarity. Then the most similar pair of sequences is aligned (using a dynamic programming algorithm). A consensus sequence is generated from this alignment, and the next most similar sequence is aligned with this consensus. Then a new consensus is calculated, and this process is repeated until all of the sequences are incorporated into the alignment. This process works fairly well, but the final alignment produced is approximate rather than optimal. Also, the sequences to be aligned must all be about the same length and have a pretty high level of similarity throughout. It is possible to detect and align small similar regions located within a group of larger sequences, but this requires a combination of pattern detection and multiple alignment algorithms.

Understanding the difference between an optimal alignment and an approximate one is less important than understanding the difference between a computationally optimal solution and a biologically meaningful one. A computational algorithm finds a maximal or minimal score for some set of rules, but these rules, however complex, always represent a simplified model of the true biological situation. Biology is full of exceptions to rules and special situations. It is often the case that the alignment produced by a computer program will need to be adjusted by hand to preserve biologically important regions—i.e., no gaps inserted in the middle of an enzyme's active site.

PATTERN FINDING

Another broad class of bioinformatics tools are used for pattern finding. Many different kinds of patterns are present in DNA and protein sequences. Some are simple, such as the DNA sites recognized by restriction endonuclease enzymes (e.g., GAATTC is recognized by the enzyme EcoRI). Others, such as conserved protein domains that fold into functional 3-dimensional structures (motifs), are complex. Therefore, the various bioinformatics tools used to detect these pattern reflect their different levels of complexity.

Simple patterns such as restriction enzyme sites in DNA can be found by an exact pattern-matching tool—just like the "Find" command in any word processing program. A slightly more sophisticated-pattern matching tool can include a mismatch at a specified location, or anywhere in the pattern. However, many biological patterns such as promoter sequences in DNA and protein functional domains require more flexible pattern-searching tools. A more complex pattern can allow for a list of different letters that can be considered to match at each position in the pattern (ambiguities) and regions of variable size where any letter is allowed: [LIVMFY]–x(2)–[STG]–G–x(2,4)–[ST]–C. In this example, dashes separate positions in the pattern, letters in brackets can be substituted at that position, x means any letter is allowed at that position, x followed by a number such as "(2)" means any two letters, x followed by a range such as "x(2,4)" means a variable number of letters is allowed (from 2 to 4).

Several databases of biological patterns have been created. Promoter sequences are stored in the Eukaryotic Promoter Database (EPD; www.epd.isb-sib.ch Perier et al., 2000), and transcription factors (a somewhat broader category, which includes enhancers and other regulatory elements that may be located some distance from the coding region of a gene) are collected in the TransFac database (transfac.gbf.de/transfac Wingender et al., 2000). Patterns for conserved protein domains can be found in **ProSite,** a dictionary of protein sites and patterns (www.expasy.ch/prosite Hoffman et al., 1999). The essential quality of a pattern is that it is hand-built by an expert biologist who has spent a lot of time scrutinizing a conserved motif in a group of related sequences. A variety of simple computer programs that can run on any PC, UNIX machine, or free Web pages are available to search any given DNA or protein sequence for matches with the appropriate set of patterns. These pattern-searching programs are fast, and they do not use much computer power, even for searches with thousands of different patterns.

PROFILE SEARCHES

Pattern searches, even with ambiguities and variable sized gaps, have some serious limitations. It is a form of exact matching, so only those variations of a real biological pattern that have been specifically included in the description of the pattern will be found. So, by definition, new sequences that have unexpected variations of this pattern will not be found. This is like not being able to find a file on your computer's hard drive with the "Find → File" utility because you don't remember how to spell the file's name. The ability of BLAST and FASTA to find *similar* sequences would be useful in pattern finding, particularly their ability to use a matrix of amino acid similarities when evaluating protein patterns.

The concept of recurring patterns is particularly well developed in the study of protein sequences. If all of the protein sequences in any of the major databases (GenBank, SwissProt, PIR, etc.) are compared to each other using a similarity tool such as BLAST or FASTA, it is immediately obvious that many of them fall into groups. Furthermore, each of these groups contains proteins with similar functions—such as kinases, methylases, and cell surface receptors—so the protein groups are actually protein families. Detailed inspection of the sequences within each protein family reveals certain regions that are highly conserved among all of the proteins in that group. Furthermore, many of these conserved regions (known as motifs) form 3-dimensional structures that play an essential role in the function of those proteins—an active site for an enzyme or a crucial protein fold that binds with a ligand or in a protein–protein interaction. Prediction of the function of new genes can be done more reliably using databases of known protein families and pattern-finding tools. A pattern-finding tool may be able to use the similarity of these motifs to identify new members of each protein family that may have too little overall similarity to be identified with BLAST or FASTA. Also, a new protein may have similarity to two or more motifs—which can provide useful information about its potential function. These multiple conserved regions would be difficult to interpret in the results of a BLAST or FASTA search.

In some cases, the structure and function of an unknown protein which is too distantly related to any protein of known structure to detect its affinity by overall sequence alignment may be identified by its possession of a particular cluster of residues classified as a motif. The motifs arise because of particular requirements of binding sites that impose very tight constraint on the evolution of portions of a protein sequence.

Lesk (1988:17)

An enhanced pattern-searching method called **profile analysis** was developed by Gribskov in 1990. A profile is a mathematical description of a conserved region, domain, or motif—usually in a protein, but DNA profiles can also be constructed—which is built from a set of sequences that all share this motif. All of the sequences are aligned (a multiple alignment), and then the frequencies of each letter are calculated for each position (the number of times each letter appears in a column of the multiple alignment). The result is a **position-specific scoring matrix** that fully describes the motif across all of the sequences used in the multiple alignment. The matrix can be filled out with zeroes for all possible amino acids (or DNA bases) that do not occur in that position in any of the sequences in the set, or the matrix can be filled out with values chosen from a table of natural amino acid mutation rates (a weighted average of the values for each of the amino acids that are in the alignment at that position). Then the matrix can be used in place of a single sequence for a modified BLAST/FASTA-type similarity search, either against a single sequence or against an entire database of sequences. Alternately, a database of profiles can be created, such as for protein domains, and a new sequence can be searched for similarity to all of these profiles.

A number of profile-based databases of protein motifs have been created. The simple patterns in ProSite have been supplemented with profiles based on multiple alignments of the conserved regions of a selected set of proteins for each of the families. This set of ProSite profiles has been expanded in the BLOCKS database (www.blocks.fhcrc.org Henikoff et al., 2000) by using the ProSite profiles to search all of the proteins in the SwissProt database for more members of each protein family. The ProDom database (prodes.toulouse.inra.fr/prodom Corpet et al., 2000) is built in a completely automatic fashion, first clustering all of the proteins in SwissProt using a similarity program (a form of BLAST) and then building a profile for each cluster (Fig. 3-5). This is surprisingly useful, since many clusters

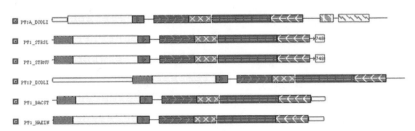

FIGURE 3-5. A set of conserved domains from the ProDom database.

are identified in ProDom that contain proteins that have no
known function (yet!) or represent small conserved domains in
groups of otherwise unrelated proteins.

HIDDEN MARKOV MODELS

The concept of using pattern-based searching for conserved
domains among related proteins makes intuitive sense. As we
learn about a family of proteins, information about a conserved
site—such as the catalytic domain for a family of enzymes with
related biochemical functions or a DNA-binding domain for a
family of regulatory proteins—should be useful for identifying
additional proteins in that family. However, this does not always turn out to be the case. A standard brute-force BLAST
search is sometimes more sensitive at discovering new members
of a protein family than is a profile-based search method. There
are a few reasons for this. First, profiles consider information
about only a small region of a protein (the conserved motif), but
there may be subtle bits of information located elsewhere in the
protein, either near the motif or distant from it, that can
contribute to the BLAST similarity score. Also, profiles evaluate
each position in a motif separately (i.e., How well does this letter
match the corresponding column of the matrix?) and ignore
interactions between different positions.

A more sophisticated pattern-search tool, known as a hidden Markov model (HMM), has been developed that takes into consideration the influence of neighboring amino acids in constructing and searching for motifs in proteins. The math behind HMMs is rather complicated, so let it suffice to say that this is a pattern-analysis technique that was developed in linguistics, has been adopted by bioinformatics, and uses a lot of computer power to make searches. The Pfam database contains about 3000 HMM profiles for protein domains (pfam.wustl.edu; Sonnhammer et al., 1998) Most of the profiles in Pfam were created from hand-built multiple alignments of conserved domains from groups of related proteins (protein families). An additional set of domains come from an automated clustering of all of the proteins in SwissProt. These profiles can then be compared to an unknown protein (or a predicted protein from a genome sequence) to identify functional domains.

Many proteins contain more than one functional domain. Sometimes each of these domains match unrelated protein families. Walter Gilbert (1978) suggested that new proteins evolve by splicing together several functional modules (motifs) from other proteins. In many cases, these motifs correspond to exons, so that recombinations that occur between introns can easily create new multidomain proteins. A similarity search with BLAST or FASTA finds regions of alignment with other proteins that have each of these functional domains and tries to extend these alignments across the entire sequence of the protein. The similarity programs rank these alignments by their overall score based on the percentage of identical and similar amino acids and the length of the aligned region between the query protein and the most similar protein in the database. This can be quite confusing when a protein shares just one domain with one protein family and another domain with another protein family. A Pfam search deals sensibly with these "shuffled chunks" of proteins by showing where each profile matches the query sequence and by giving a score for how well just that region matches the profile of that conserved domain. HMM searches with Pfam profiles have been helpful in the initial stages of annotating all of the new proteins discovered in whole genome sequencing projects, including the human genome.

PHYLOGENETICS

Databases of protein domains and motifs are created from clusters of proteins that contain regions of similar sequence. It is important to distinguish between a similarity based on true homology, which is the result of genes that share a common ancestor, as opposed to a strictly functional similarity based on protein regions that contain a disproportionate amount of one or two amino acids (e.g., a proline-rich region or a membrane-spanning region that contains many nonpolar amino acids). It can also be confusing to look at a family of proteins that share a common domain and see several proteins from a single species and a number of others from different species. To make some sense out of the relationships between the members of a protein family, it is necessary to understand something about the process of evolution at a molecular level.

New genes are created by two distinctly different processes: gene duplication and speciation. Genes (or entire chromosomes) can be duplicated in a number of different ways during the processes of replication and recombination. Once a species contains two copies of a gene, random mutation events cause independent changes in the two sequences. These mutations may lead to one gene copy taking on new a function while the other copy provides the original function. These two gene copies are known as **paralogs.** Alternately, if two populations of a species are separated for a long time, each population will accumulate different mutations in its genes until the two populations form two distinctly different species. Now each gene in species A has a similar but not identical match in species B. These are known as **orthologs.**

Gene families are created by complex combinations of gene duplication and speciation, so it is not always obvious when looking at two gene sequences from two species whether they are orthologs or paralogs. This is extremely important if we are going to rely on comparisons between species to assign functions to genes—such as mouse knock-out experiments to explore the function of corresponding human genes. The only reliable method for determining the relationships between the members of a multigene family is to calculate an evolutionary tree. Figure 3-6 shows a tree containing the family of paralogous globin genes in humans. There is an entire branch of biology called taxonomy, which is devoted to defining the evolutionary relationships between species; and the majority of work in this field in the past few decades has focused on methods for the use of DNA and protein sequence data.

Scientists working in numerical taxonomy or phylogenetics have created some complex mathematical approaches to

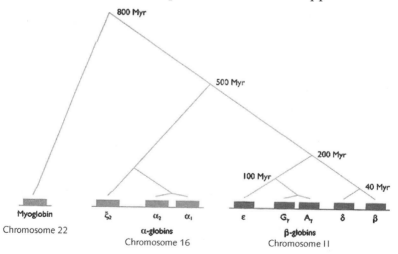

FIGURE 3-6. The globin gene family has been created by gene duplications. Reproduced, with permission, from T. Brown, *Genomes* 2nd edn. Copyright 2002, BIOS Scientific Publishers Ltd.

calculating the relationships among gene sequences, which goes far beyond the scope of this book. However, unlike the work on sequence similarity, multiple alignment, or protein domains, there are no broadly accepted phylogenetic methods that consistently achieve good results. Some methods focus on grouping sequences based on their absolute numbers of similar and different bases (or amino acids for proteins). This creates a good quantitative estimate of current similarities but ignores evolutionary processes that may have made large changes to some genes in short periods of time. Other methods focus on recreating the exact process of mutations from a common ancestor to create the current set of observed gene sequences; but without data from all of the intermediate forms (many of which are presumably extinct), there can never be an absolutely accurate calculation. None of these methods can adequately account for the messy realities of evolution, which may include the mixing of different genes by recombination, hybridization between different species, and other phenomena that don't follow clean mathematical models.

Despite these theoretical hazards, there are a variety of phylogenetic software tools that can be used to define the relationships among clusters of genes in closely related species with reasonable reliability. The basis of all phylogenetic calculations is a multiple alignment of the relevant genes. Within a multiple alignment, pairwise similarities are calculated and a tree is built from the branches back to the base by joining the most similar pairs, then the next most similar, etc. In general, orthologs cluster together most tightly unless there has been a recent gene duplication event that affected one species but not others.

REFERENCES

Altschul SF, Gish W, Miller W, et al. Basic local alignment search tool. J Mol
    Biol 1990;215:403–410.

Altschul SF, Madden TL, Schäffer AA, et al. Gapped BLAST and PSI-BLAST: A new generation of protein database search programs. Nucleic Acids Res 1997;25:3389–3402.

Corpet F, Servant F, Gouzy J, Khan D. ProDom and ProDom-CG: Tools for protein domain analysis and whole genome comparisons. Nucl Acids Res 2000;28:267–269.

Dayhoff MO, Schwartz RM, Orcutt BC. A model of evolutionary change in proteins, matrixes for detecting distant relationships. In MO Dayhoff, ed. Atlas of protein sequence and structure. Vol. 5. National Biomedical Research Foundation. Washington, DC, 1978;345–358.

Feng DF, Doolittle RF. Progressive sequence alignment as a prerequisite to correct phylogenetic trees. J Mol Evol 1987;25:351–360.

Gilbert W. Why genes in pieces? Nature 1978;271:501.

Gribskov M, Luethy R, Eisenberg D. Profile analysis. Meth Enzymol 1990;183:146–159.

Henikoff JG, Greene EA Pietrokovski S, Henikoff S. Increased coverage of protein families with the Blocks database servers. Nucl Acids Res 2000;28:228–230.

Hofmann K, Bucher P, Falquet L, Bairoch A. The PROSITE database, its status in 1999. Nucleic Acids Res 1999;27:215–219.

Lesk AM. Computational molecular biology. Oxford, UK: Oxford University Press, 1988;272 pp.

Needelman SB, Wunch CD. A general method applicable to the search for similarities in the amino acid sequence of two proteins. J Mol Biol 1970;48:444–453.

Pearson WR. Rapid and sensitive sequence comparison with FASTP and FASTA. Meth Enz 1990;183:63–98.

Pearson WR, Lipman DJ. Improved tools for biological sequence comparison. Proc Natl Acad Sci USA 1988;85:2444–2448.

Perier RC, Junier, T, Bonnard C, Boucher P. The Eukaryotic Promoter Database (EPD). Nucl Acids Res 2000;28:302–303.

Smith TF, Waterman MS. Identification of common molecular subsequences. J Mol Biol 1981;147:195–197.

Sonnhammer EL, Eddy SR, Birney E, et al. Pfam: Multiple sequence alignments and HMM-profiles of protein domains. Nucl Acids Res 1998;26:320–322.

Wingender E, Chen X, Hehl R, et al. TRANSFAC: An intergrated system for gene expression regulation. Nucl Acids Res 2000;28:316–319.

GENOME DATABASES

WHAT IS A GENE?

The goal of medical genomics is to use the information generated by automated sequencing technologies to find the genes
that will enable new methods to diagnose and cure disease. Most
biologists and medical professionals have a general concept of a
gene as a region of DNA on a chromosome that encodes a
protein. They also recognize that genes come in different allelic
forms and that these alleles segregate and combine in predict-
able dominant–recessive pairs to determine the characteristics of
the progeny from a given pair of parents.

Modern molecular genetics is still more or less based on the
Beadle and Tatum (1941) One Gene = One Protein model. However, additional discoveries have greatly complicated this neat
model. Genes code for many different kinds of proteins in
addition to enzymes, and some functional proteins are composed of the products of multiple genes (multiple subunits).
Some genes do not code for proteins at all, but are transcribed
into functional RNA molecules, such as ribosomal RNA (rRNA)
and transfer RNA (tRNA). For genes that do code for proteins,
the mRNA is actually subjected to many modifications before it

Essentials of Medical Genomics, Edited by Stuart M. Brown.
ISBN 0-471-21003-X.   Copyright © 2003 by Wiley-Liss, Inc.

is translated. Prokaryotic genes may be transcribed in multigene units from a single promoter (operons) and are then cut into individual mRNAs. Eukaryotic genes are usually interrupted by nonprotein coding sequences, known as **introns,** which must be spliced out of the mRNA before the message is translated into protein. Making this even more complicated, some mRNAs can be spliced in several different ways to produce distinctly different templates and, therefore, different protein products (**alternative splicing**).

The **promoters** and **transcription factors,** that determine when and how much of each gene will be transcribed into RNA, are not located in the part of the gene that is copied into RNA (the coding region), yet they are essential parts of each gene. Furthermore, each mRNA is not translated into protein in its entirety but contains some leader sequence before the protein-coding region and some additional "downstream" sequence after the end of transcription. These non-translated sequences on the mRNA have important regulatory roles that help determine the amount, timing, and tissue specificity of protein production from each gene.

The current result of all of this additional complexity in our understanding of molecular biology is that no two biologists can agree on a completely consistent definition of a gene. It is more than a region of a chromosome that encodes a protein, because there are genes that do not code for proteins, and there are important regulatory parts of genes that are outside the coding region. It is not the DNA instructions for producing a specific protein, for many different proteins can be produced from one gene due to alternative splicing and posttranslational modification of proteins.

The lack of a clear definition of a *gene* is a major stumbling block in reaching the goal of defining a comprehensive parts list for human beings in terms of DNA and protein sequences. This is clearly the essential next step in the Human Genome Project

(HGP), which is being eagerly awaited by both clinical and basic scientists. Once we have a list of all the genes or all the proteins, then we can start to apply high-throughput technologies, such as microarrays and protein–protein interactions with some degree of quantitative precision. Without such comprehensive lists, there is still room for murky, undefined factors in any biological experiment on humans. Thus the focus of attention in the HGP and the genomics community has shifted from automated DNA sequencing technology (including the assembly of reads into contigs and contigs into whole chromosomes) to genome annotation and genome databases.

GENBANK

Now that the HGP is nearly complete, how can ordinary people look at the data or make use of it? All of the data from the HGP are available to anyone with a computer connected to the Internet. The National Center for Biotechnology Information (NCBI) a branch of the National Library of Medicine at the National Institutes of Health (NIH) maintains a database known as GenBank, which is the complete definitive public collection of DNA sequence information. There are some private databases owned by companies such as Celera Genomics and Incyte; for the most part, however, they contain the same data but in more depth (more copies of overlapping clones that allow for more accurate consensus sequences and fewer gaps) and with alternate computer predictions of gene coding regions.

GenBank is most easily accessed at the NCBI web site (www.ncbi.nlm.nih.gov). This Web site is the entry point to a very sophisticated set of searching tools. There is a tool called **Entrez,** which performs text and keyword searches on all of the annotations in the DNA and protein database files. Queries can be made for gene names, diseases, and individual scientists who submitted sequences, and they can be limited by organism, chromosome, date, etc. Protein sequences are linked to their corresponding DNA sequences and to locations on the genome (when known). In addition, Entrez is linked to **MEDLINE (PubMed),** the comprehensive database of all medical and scientific literature maintained by the National Library of Medicine. Again, every protein and DNA sequence is linked to relevant journal articles, and vice versa. Links are also maintained to protein 3-dimensional structures and human genetic diseases, creating a rich set of relationships across all of these interrelated databases (Fig. 4-1).

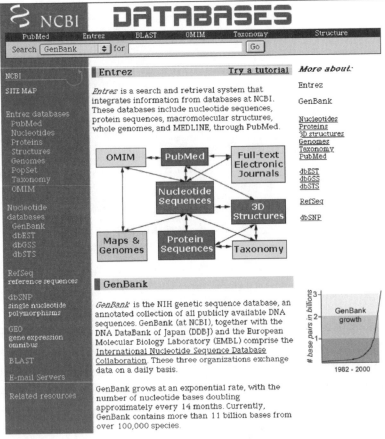

FIGURE 4-1. An overview of the databases available on the NCBI Web site.

The NCBI also maintains a tool called BLAST (Basic Local Alignment Search Tool) which searches the DNA and protein databases by making comparisons between a sequence supplied by the user and all of the database sequences (see Chapter 3). It is also possible to compare DNA to protein sequences using automatic translation. Considering that there are many millions of sequences in GenBank (billions of bases of DNA), the NCBI's BLAST server is amazingly fast. The proper interpretation of the results of a BLAST search requires a thorough understanding of molecular biology, evolution, and statistics, so it is not really intended for use by the general public. However, the tools are free for anyone to use (www.ncbi.nlm.nig.gov/blast) (Fig. 4-2).

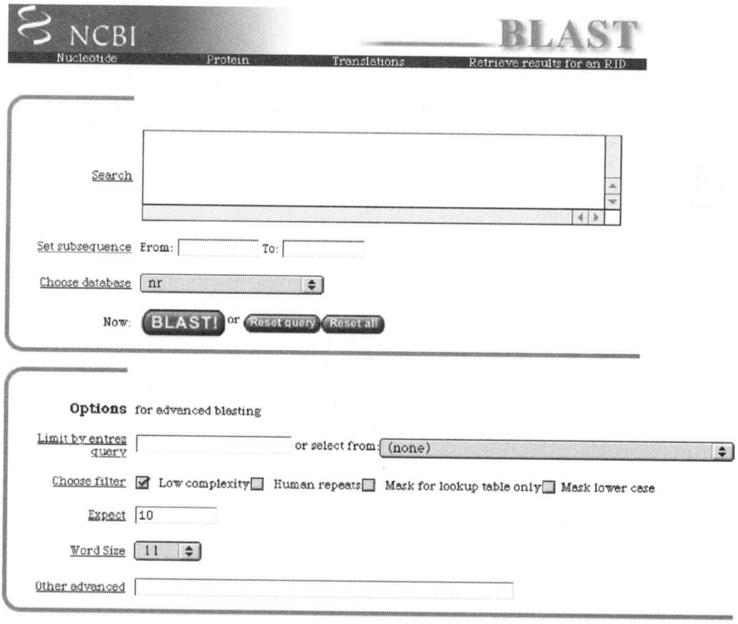

FIGURE 4-2. The BLAST query page on the NCBI Web site for DNA–DNA searches.

GENOME ANNOTATION

The initial products of genome sequencing are millions of bases of DNA—just an endless stretch of seemingly random G, A, T, and C letters (Fig. 4-3). The sequence is only meaningful when the genes are found and their biological functions are described. Ideally, we would like to have a genome that is deeply annotated with multiple layers of relevant information. We would like to know the locations of all protein-coding regions, intron–exon structure, alternative splicing sites (and the corresponding transcripts), promoters and transcription factor–binding sites, protein structure and function, including roles in metabolic and regulatory pathways, protein–protein interactions, and gene expression data (which might include information from a wide variety of healthy and diseased tissue types and responses to various drugs and environmental perturbations). In a sense, the annotated genome should become a central tool for organizing knowledge about genes.

All of this annotation rests fundamentally on the ability to find genes in the genome sequence. That turns out to be a challenging problem, partly because of the difficulty in defining

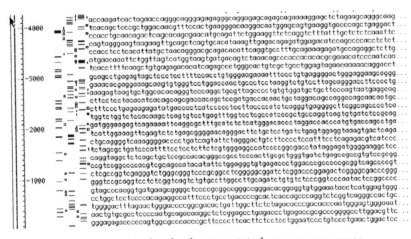

FIGURE 4-3. A chunk of unannotated genome sequence.

what is a gene in a manner that satisfies the majority of biologists and that can be specified in unambiguous computer code and partly because the genome is full of sequences that look like genes but are not (pseudo-genes) and because some genes are hidden. Another, more fundamental, problem is in the basic concept of pattern recognition. We know the sequences of some genes, so we look for other genes that resemble the ones that we know. This similarity may be based on direct sequence–sequence matching (i.e., BLAST), or it may rely on more subtle statistical properties of the DNA that can be found to differentiate gene-coding regions from noncoding regions. Either way, we are relying on what we know to shape what we are looking for, so it is inevitable that we will miss things that are novel. Yet biology is always full of exceptions to the rules and new rules that govern new subclasses that can be discovered only by using fundamentally different methods.

The genome sequencing projects have come to depend heavily on comparisons of genome sequences against databases of "known" sequences using a similarity tool. Yet BLAST is a rather simple-minded, brute-force algorithm that does not take into account the deep complexity of biological systems. Each chunk of genome sequence is compared to all of the known genes in **GenBank** (which have been previously described by molecular biologists) using BLAST. As the BLAST matches are identified, the genome sequence can be annotated with the location of genes, and the GenBank genes can be cross-referenced to a specific spot on a chromosome. By locating each gene on the genome, flanking sequences can be defined—these play an important role in the regulation of gene expression. One of the primary goals of the HGP is to create a complete listing of all human genes and their locations on the genome.

The process of identifying genes in the genome using BLAST has many problems. GenBank does not yet contain a comprehensive list of all human genes and their functions. As of late

2001, the RefSeq database at NCBI contained about 14,500 putative human genes that are thought to have unique full-length mRNA sequences and some functional information. Yet when these mRNAs were aligned with the genome sequence (using BLAST and similar tools), some failed to match, some matched in more than one location, and there were some cases of multiple RefSeq mRNAs matching a single spot on the genome (Lander et al., 2001). The SwissProt database has somewhat stricter criteria before it adds a sequence; and in early 2001, it only listed about 5,400 human proteins. On the other hand, the protein section of GenBank (GenPept) contains about 78,000 human (and other primate) proteins—but there has been no attempt to make this a nonredundant set.

MATCHING ESTs TO THE GENOME

Much of the current progress in the annotation of the human genome involves using BLAST to match random fragments of mRNA sequence from various expressed sequence tag (EST) databases to the genome sequence. ESTs are sets of mRNAs that are collected from specific tissues (fetal brain, liver tumor, etc.), copied into complementary DNA (**cDNA**) with the enzyme reverse transcriptase, and cloned into plasmids. Then a single sequencing read is taken from each clone—so this is a sequence tag rather than the complete sequence of the mRNA. It is possible to collect many thousands of EST sequences from a single tissue quickly and inexpensively, producing an expression profile of the genes being transcribed in that tissue. In theory, if enough ESTs are collected from enough different tissue types under various developmental, disease, and environmental conditions, a complete expression profile could be produced for the entire genome of an organism. GenBank contains >3.8 million human ESTs (as of November 2001), but EST collections are never complete. The mRNAs for some genes are more common than others. In fact, in a random collection of a few thousand cDNA sequences cloned from a particular tissue, more than half will be from just a few dozen genes. Some genes, such as regulatory elements, are transcribed at very low levels, whereas others may only be expressed at a precise point in development or under unusual conditions. Therefore, EST collections will never contain absolutely every single gene.

The human genome contains many duplicated genes (paralogs) that may have slightly different coding sequences, different introns, and certainly different flanking sequences—so it is nearly impossible to make unambiguous matches for each mRNA sequence back to its gene of origin. There are also pseudogenes—copies of genes that are no longer functional, and it is often difficult to identify these on the basis of a BLAST match. Human genes are also alternatively spliced, so that a single gene can be the template for the production of many different mRNA molecules, leading to multiple ESTs that are produced from the same genomic region, yet lack sequence similarity to each other.

Another problem with using similarity-based methods to find and annotate genes in the genome is the old "garbage in, garbage out" problem. A lot of the EST sequences (and even the "official" GenBank sequences that include a coding region and a protein translation) have no useful functional annotation. A lot of these are anonymous sequences, which may be annotated with the description "hypothetical protein" or with a description of the tissue from which they were cloned. Many of the entries in GenBank are now submitted in huge batches directly from a computer at a sequencing center to another computer at GenBank with no human involved in the process at all. The description fields of these database entries are filled in automatically—generally by making a BLAST search of GenBank. The net result can be that a new sequence is annotated with information like this: "Similar to hypothetical protein A200456." In other words, nonsense is propagated from one unknown sequence to another one that is similar to it. The main utility of EST and other cDNA sequence databases for genome scientists is to confirm computer-predicted genes in genomic sequences based on similarity matches to cDNA (although non-protein-coding genomic DNA fragments have been found in what were supposed to be databases of cDNA sequences).

THE ANNOTATED GENOME

The NCBI provides a graphical map of the human genome, which shows the location of all known genes, ESTs, cytogenetic bands, and many other genetic markers (Fig. 4-4). It is now routine for scientists to isolate a piece of RNA or DNA in a laboratory experiment, sequence it, and then find its exact location on the genome using BLAST and the genome map. The complete genome sequence also provides information about introns, promoters and other flanking sequences, nearby genes, etc.

OTHER VIEWS OF THE HUMAN GENOME SEQUENCE

There are two other full-featured genome maps available for free on the Web. The University of California at Santa Cruz (USCS) provides the Golden Path genome browser (genome.ucsc.edu Kent et al., 2002). Jim Kent at UCSC has built a superior tool for assembling large chunks of genome data (contigs) into a continuous "Golden Path" through entire chromosomes. The UCSC map viewer's data display is superior to that of the NCBI genome map in many respects, but it does not include the powerful Entrez search engine for locating genes by complex queries based on name and/or keywords (Fig. 4-5). Also, not all GenBank sequences are included in the UCSC map.

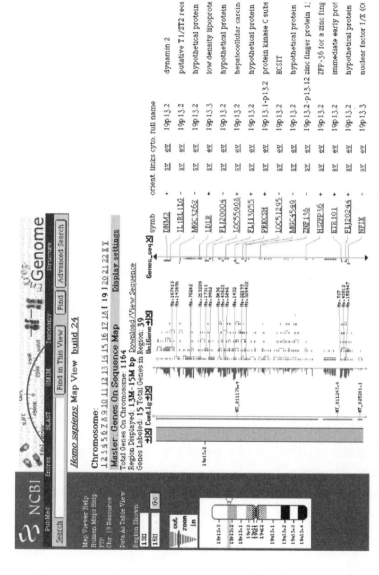

FIGURE 4-4. A section of chromosome 19 in the NCBI's human genome map.

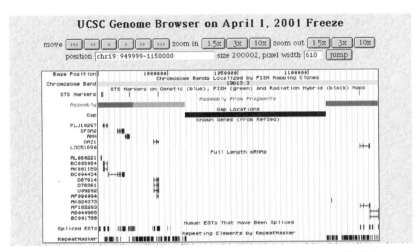

FIGURE 4-5. A segment of chromosome 19 using the UCSC human genome browser. Reprinted with permission: Univ. California, Santa Cruz Genome Bioinformatics Human Genome Browser; genome.ucsc.edu.

The European Molecular Biology Laboratory (EMBL), the European counterpart of the NCBI, created a tool called **Ensembl** that combines features from Entrez, BLAST, the NCBI genome map, and the UCSC genome browser. Although Ensembl has its flaws and idiosyncrasies, it provides direct links from every position on its map to the equivalent position on the NCBI and UCSC maps.

INCONSISTENCIES IN THE GENOME DATA

The public availability of human genome data and fairly nice graphical browsers may suggest that the genome sequence is all falling neatly into place. The reality is far more complex and confusing. Alternate splicing leads to many different mRNAs that map to a single genomic locus, yet these different mRNAs may be translated into different proteins that have different biological functions and have different expression patterns. Some alternately spliced forms of mRNA may be predicted by computer models but never found in tissues, while real cDNA sequences may include splice variants that are not predicted by computer models. In fact, real cDNAs may include exons so far upstream or downstream from the rest of the gene that they overlap with exons of other genes (Fig. 4-6).

The current assemblies of human genomic data are far from perfect. In fact, a large percentage of the genome sequence contains duplicated regions due to incorrect assembly of contigs and incorrect overlapping of subclones. In addition, there are

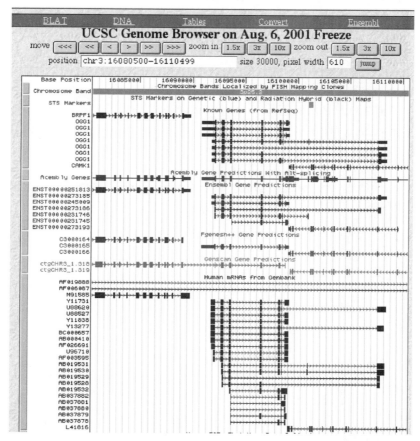

FIGURE 4-6. A region of chromosome 3 using the UCSC human genome browser, showing multiple mRNAs mapping to a single locus and alternately spliced forms of two genes with overlapping exons. Reprinted with permission: UCSC Genome Bioinformatics Human Genome Browser; genome.ucsc.edu.

real genomic duplications that lead to paralagous genes (genes with similar coding sequence) or pseudogenes (a copy of a gene located elsewhere in the genome that does not produce a functional cDNA transcript). Therefore a cDNA may match several different genomic locations, which potentially could lead to several different flanking sequences, making it difficult to identify common promoter elements in gene expression studies.

HUMAN GENETIC DISEASES

Rather than start a genome search with a piece of DNA sequence or a complex query in the scientific literature, a physician (or a patient) is more likely to start with a known disease. The NCBI has built a companion to the GenBank/PubMed database—called Online Mendelian Inheritance in Man (OMIM)—that is completely focused on human genetic diseases (Fig. 4-7). OMIM

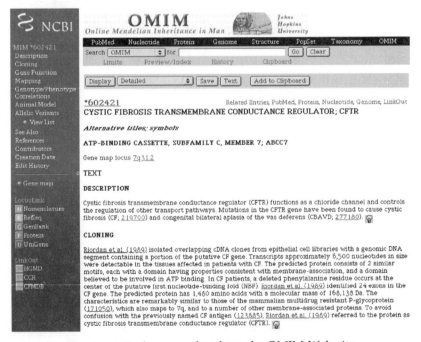

FIGURE 4-7. A screen shot from the OMIM Web site.

is authored and edited by Dr. Victor McKusick and co-workers at Johns Hopkins University. OMIM contains a short description of each gene and extensive excerpts and summaries of a wide range of scientific literature on the gene and the disease, including clinical reports, all known alleles and mutations, an extensive bibliography (with direct links to PubMed citations for each paper), links to the GenBank entries for both the gene and the protein, links to a cytogenetic map (which is, in turn, linked to the NCBI's human genome map and the complete genomic sequence, neighboring genes, and known mutations in that region). OMIM entries also have links to other disease-specific databases that have relevant information.

A System for Naming Genes

Unfortunately, searching genome databases is not always as simple as typing in the name of a disease or gene and hitting a "Search" button. The problem is not inherent in the structure of computer databases or even in the complexity of biology but simply that people do not always call the same thing by the same name. In my own work as a bioinformatics consultant, I often encounter scientists in a particular discipline—be it immunology or microbiology—who call a particular gene by particular name, but a search on that name comes up empty in GenBank. It seems that the database curators have chosen another name for that gene. The only commonality is the sequence itself and the accession number. Another problem is the rampant naming of genes by geneticists. People working with fruit flies like to name genes after the appearance of flies with a mutation in that gene. However, mice or humans with a mutation in the corresponding (homologous) gene will not have the same appearance (wingless, bent antennae, etc.), so that gene will have a different name in each species. Later on, if a similar gene were to be found in a worm or a plant, would it then be described as similar to the fly gene or similar to the human gene?

This problem has become worse and worse as more genes have been described in more species and studied in different contexts. Fortunately, a group of geneticists and database curators have recently begun a project called the Genome Ontology (GO) to sort out all of the names into a consistent system. The names of all organisms on earth have been organized into a single taxonomy, so that with a single term a scientist can communicate unambiguous information about the identity of an organism and, at the same time, indicate the relationship of that organism to all other living things. A gene ontology would include both unambiguous names for all genes as well as a consistent vocabulary to describe the features of genes that are not specific to any one type of organism or any particular scientific discipline (at the expense of others).

GO is organized around three general principles that are common to all eukaryotic organisms: molecular function, biological process, and cellular location. In addition to developing an internally consistent vocabulary that can apply equally well to all organisms and biological disciplines, the GO project has taken on the task of reannotating all of the existing gene, protein, and species databases by mapping all of the terms used in current gene feature descriptions to equivalent GO terms. Once these equivalencies are established, then GO can serve as an intermediary to translate the annotation terms of any database to any other database. Then if scientists can learn to phrase their database queries using GO terms, the same query will work equally well in all databases.

MODEL ORGANISMS (COMPARATIVE GENOMICS)

This discussion of ontology across species brings up an interesting question. Are there equivalent genes for all functions in all organisms? Intuitively, one would say no; after all, different organisms have different amounts of DNA and different numbers of genes. On the other hand, all organisms have more or less the same biochemical processes at the cellular level—energy metabolism, growth, reproduction, movement, etc. At the level of protein sequences, quite a lot of similarity (homology) can be found across distant branches of the tree of life. Yet some groups of organisms have unique structures or unique metabolic processes. Within the group of mammals, there appears to be a common set of genes, some of which may be duplicated or lost in any particular species; but overall, **orthologs** (homologous genes with identical function in different species) can be found for almost every gene.

So how relevant are gene homologies among various model organisms to the practical aspects of medicine? First, remember that most drugs are tested on animals before they are taken to clinical trials on humans, both for reasons of safety and to have better control of the experiments. There are many reasons why basic research relies on animal models, but it all boils down to this: The human is a poor experimental subject. In humans, it is not possible to make controlled mutants or gene knockouts or to make controlled breeding experiments. There is not even a comprehensive collection of mutants. In contrast, the mouse is an almost perfect experimental subject. It is small and can easily be grown in the laboratory. It has a short generation time and is highly prolific. There are thousands of pure-bred strains that contain individual, well-characterized mutations; and any individual gene can be knocked out using standard procedures.

Animal models, particularly the mouse, are going to be extremely important in the next phase of annotating the human genome. First, in discovering all of the genes and, then, in defining their functions in increasingly fine detail. In mid-2001, the mouse genome was at the forefront of the genome-sequencing race. Celera Genomics officially completed their version of the mouse genome sequence, when the public genome project was still >1 year away from completion. The HGP passed this critical official completion mark and is now in the much-slower and less-exciting phase of fixing errors and improving the annotation, which will take years. UCSC now has a mouse genome browser that shows homologies to human genes.

HUMAN–MOUSE SYNTENY

It is interesting that among related organisms, such as mice and humans, not only are there orthologs but there are extensive sections of chromosomes that contain these similar genes in the exact same order. In fact there are a few hundred blocks of **synteny** (conserved gene order) that account for a huge fraction of the mouse and human genomes. In other words, all the genes on the human chromosomes can essentially be reorganized into the order that they are present on the mouse chromosomes with a few hundred cut-and-paste operations. The tens of thousands of genes have not been randomly shuffled by all of the millions of recombination events that have occurred since evolution separated the two species. For example, human chromosome 19 has about 1200 genes, which are found in essentially the same order in 15 segments spread across the mouse chromosomes. (Dehal et al., 2001) (Fig. 4-8).

These human–mouse syntenic segments are the focus of a lot of activity among genome scientists. Where a similar gene is found in both mouse and human, gene function can be investigated in the mouse by making a gene knockout. In some cases, a mouse or a human gene is known from a cDNA database but no corresponding gene can be found in the other species by a simple homology search. The corresponding gene can be found in the other species by defining known genes on either side and then looking for open reading frames (**ORFs**) in the corresponding region of genomic DNA. When syntenic segments of mouse

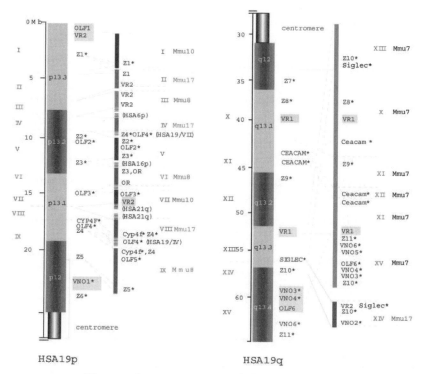

FIGURE 4-8. Human chromosome 19 and the 15 corresponding syntenic segments of mouse chromosomes. (Reprinted with permission, Science Magazine, from Dehal et al., 2001.)

and human genomic DNA are aligned, protein-coding sequences show much greater sequence similarity than noncoding regions. It is possible to identify new genes that are not known in the human or mouse simply by finding regions within aligned syntenic segments that show increased sequence similarity (and that have the potential to code for protein).

It is also interesting to investigate the regions between genes in syntenic segments of human and mouse genomic DNA. Since this is non-protein coding sequence, millions of years of accumulated mutations should have obliterated any similarity. However islands of similar sequence still exist – which may correspond to regulatory regions. The Vista website (www.???) provides a graphical browser that shows the level of similarity within and between genes in aligned human and mouse genomic segments.

The comparison of genome sequences from human and mouse has led to a number of fascinating evolutionary insights. Essentially, all of the single-copy genes are equivalent between human and mouse, with protein sequences that are generally 90–95% identical and located in the same order on their respective chromosomes. However, some genes are tandemly duplicated in one or both genomes, leading to families of similar genes. About 30% of human and mouse genes are members of these tandemly duplicated gene clusters. Syntenic gene clusters between human and mouse contain different numbers of genes and appear to be the result of differences in the founder genes that were duplicated, differential gene loss, and independent selection in each cluster since the divergence of primate and rodent lineages. A complex pattern of lineage-specific gene duplication and loss is evident. Some gene copies may become inactivated by mutations (pseudogenes) and others may develop unique tissue-specific or developmentally regulated gene expression patterns. Many of the breaks between the 15 syntenic segments of human chromosome 19 occur in the middle of the tandemly duplicated clusters, so these clusters may also play a key role in large-scale genome evolution.

There are some online databases that have been established for exploring the relationship between the mouse and human genomes—both for individual genes and for entire chromosomes. The NCBI has a nice human–mouse homology map (www.ncbi.nlm.nih.gov/Homology). It can be anchored on the human genome map (with cytogenetic positions) to show the corresponding mouse chromosomal segments or to show matching human segments on the mouse chromosome map. Known genes are indicated by name, and they are linked to descriptions in the LocusLink (www.ncbi.nlm.nih.gov/Locuslink) database. The presence of sequence tagged sites (STSs) are also noted (Fig. 4-9).

Human STS	Cytogen Pos	Human Symbol	Mouse chr	Mouse Symbol	cM position	Mouse STS
•	19p13.1-p13.2	PRKCSH	9	Prkcsh	6	•
••	19p13.2	ELAVL3	9	Elavl3	5	
••	19p13.2p13.1	CNN1	9	Cnn1		•
••	19p13.3p13.2	ACP5	9	Acp5	6	•
••	19q13.3	ASNA1		Asna1		
••	19p13.2	JUNB	8	Junb	38.6	
••	19p13.2	DNASE2	8	Dnase2	38.6	
•	19p13.13-p13.12	KLF1	8	Klf1	38.6	
••	19p13.2	FARSL	8	Farsl		
•	19p13.3-p13.2	CALR	8	Calr	37	
••	19p13.2	RAD23A	8	Rad23a		
••	19p13.2-p13.1	CACNAIA	8	Cacnala	38.5	
••	19	ETR101	8	Ier2	38.4	•
••	19p13.3	NFIX	8	Nfix	38.6	
•	19p13.1	RFX1	8	Rfx1	38	
•	19p13.11	WSX-1	8	Wsx1-pending		
••	19	EEF1D	8	5730529A16Rik		•
•	19p13.2	RPS28	8	Rps28		
	19p13.2	LYL1*	8	Lyll	38.5	
	19p13.2	KIAA0973*	8	Snsi-pending	38.6	
••	19p13.2	GCDH*	8	Godh	38.6	•••
••	19p13	CD97	8	Cd97	38	•
••	19	DDXL	8	2610307C23Rik		
•	19p13.1-p12	PRKCL1	8	Prkell	38	
•	19p13.1	PTGER1	8	Ptgerl	38	••
••	19p13.1	C190RF3		Rgs19		
•	19p13.2	DNAJB1		Dnajb1		
•	19p13.12-p13.11	NDUFB7		1110002H15Rik		•
•	19	GPSN2		A230102P12Rik		•
	19p13.1	OR7C1		Olfr57		
••	19ptcr-p13.3	SLC1A6	10	Slcla6	42.3	
••	19ptcr-p13.3	NAKAP95	10	Nakap95-pending		
••	19p13.1-q.12	AKAP8	10	Akap8		•
•	19p13.1	BRD4	10	Brd4		

FIGURE 4-9. A section of human chromosome 19 and its syntenic mouse segments using the NCBI human–mouse homology viewer.

The Jackson Laboratory, a large-scale producer of genetically pure strains of mice, also has a mouse–human genetic map on its web site (www.informatics.jax.org). This map is part of the Mouse Genome Database (Blake et al., 2001), so it is organized around mouse genetic information rather than human (Fig. 4-10).

The Sanger Centre (headquarters of the UK genome-sequencing effort) has produced a nice synteny map of human chromosome 22 and the mouse genome that shows that just eight syntenic segments span all of chromosome 22 (Fig. 4-11).

SEQUENCING OTHER GENOMES

The sequencing of the human genome has received the most attention, but the genomes of many other organisms are also

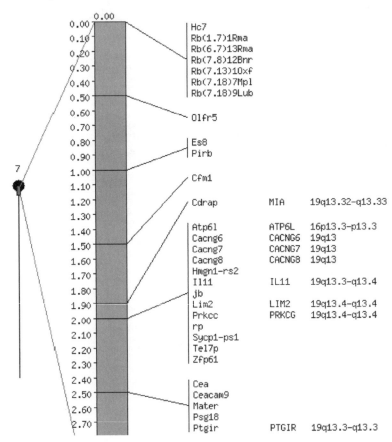

FIGURE 4-10. The Mouse Genome Database at the Jackson Laboratory. Reprinted with permission from the Mouse Genome Informatics web site, The Jackson Laboratory, Bar Harbor, Maine; www.informatics.jax.org.

medically important. Many human diseases are caused by pathogens, parasites, and their vectors. Complete genomic sequences have been available for several years for many bacteria. In fact, the first nonviral genome to be fully sequenced was *Haemophilus influenzae* in 1995 (Fleischmann et al., 1995).

As of mid-2001, there were 58 complete bacterial genomes in GenBank (11 Archaea and 47 Eubacteria), including *Chlamydia trachomatis, Helicobacter pylori, Staphylococcus aureus, Streptococcus*

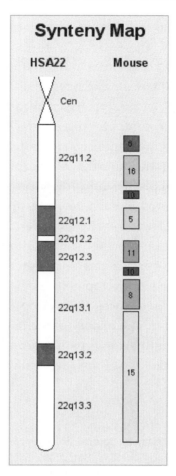

FIGURE 4-11. The Sanger Centre synteny map of human chromosome 22 and the mouse genome. Reprinted with permission from the Wellcome Trust Sanger Institute website: www.sanger.ac.uk/HGP/chr22/mouse.

pneumoniae, and *Vibrio cholerae.* The Institute for Genomic Research (TIGR) completes several more each month. Many pharmaceutical companies have in-house sequencing projects and private databases of genomic sequences for many other pathogenic microorganisms. Genomic sequences are also being determined for many eukaryotic human pathogens such as

Plasmodium falciparum, Leishmania spp., *Pneumocystis* spp., and *Giardia* spp.

In theory, protein sequences found in these genomes can be used as targets for the design of new drugs or for the development of vaccines. However, this has proven to be more difficult than expected. While computer algorithms for the detection of genes in prokaryotic genomic sequences have been quite successful, determining the function of the proteins encoded by those genes has been problematic. It is surprising that about 40% of the genes discovered in each new organism have no known role in cell metabolism or physiology. Even the identification of immunologically active proteins, such as those that are excreted or present at the cell surface, has been only moderately successful as an aid in vaccine development. Perhaps expectations for the use of genomic sequences to develop drugs and vaccines should be tempered by the history of the battle against viruses. The genome of the influenza virus has been known since 1982, yet millions of people are still infected with flu each year.

REFERENCES

Beadle GW, Tatum EL. Genetic control of biochemical reactions in *Neurospora*. Proc Natl Acad Sci USA 1941;27:499–506.

Blake JA, Eppig JT, Richardson JE, et al. The Mouse Genome Database (MGD): Integration nexus for the laboratory mouse. Nucleic Acids Res 2001;29:91–94.

Dehal P, Predki P, Olsen AS, et al. Human chromosome 19 and related regions in mouse: Conservative and lineage-specific evolution. Science 2001;293:104–111.

Fleischmann RD, Adams MD, White O, et al. Whole-genome random sequencing and assembly of *Haemophilus influenzae* Rd. Science 1995;269:496–512.

Kent JW, Sagnet CW, Furey TS. The human genome browset at UCSC. Genome Res 2002;12:996–1006.

Lander ES, Linton LM, Birren B, et al. Initial sequencing and analysis of the human genome. Nature 2001;409:860–921.

HUMAN GENETIC VARIATION

Much of what makes us unique individuals is that the DNA sequence in each of us is different from that of other people. The data from the Human Genome Project (HGP) indicate that on average any two people have 99.9% identical DNA sequences. Yet that 0.1% difference is spread over 3.2 billion bases of DNA and thus amounts to a significant number of distinct genetic traits that uniquely distinguish the genome of every person. In fact, the HGP now estimates that there are just 32,000 functional genes in the human genome. For each of these genes, there exist many different variant forms (known as **alleles**) in the human population, and each person has a unique combination of these forms.

MUTATION

Many heritable diseases are caused by a defect of a single gene. Most of these genetic diseases are recessive—a mutation has created a dysfunctional allele that does not produce an essential protein; therefore, a person who inherits two such mutant alleles cannot make the protein. For some genetic diseases, a particular

Essentials of Medical Genomics, Edited by Stuart M. Brown.
ISBN 0-471-21003-X.   Copyright © 2003 by Wiley-Liss, Inc.

mutant allele exists at an elevated frequency within a genetically isolated population (e.g., Tay-Sachs), while other diseases can be caused by any combination of a number of different mutant alleles of the gene (e.g., phenylketonuria). Other heritable diseases, such as specific forms of cancer and heart disease, are caused by a combination of alleles from number of different genes—either in an additive fashion or as a specific combination of interacting alleles. Taken by itself, each allele that contributes to a multigene disease may not be an obviously dysfunctional mutation, but rather a variant allele that is present in the population in some measurable frequency.

In evaluating the medical significance of these variable genetic traits, it is important to keep in mind the mechanisms by which they arise and are spread through populations. DNA is an extremely stable molecule, and the cellular machinery that copies DNA during the processes of growth and sexual reproduction work with good fidelity. However, errors do creep in, and these are called **mutations.** External factors, such as radiation and chemical mutagens, can damage the DNA molecule, which may lead to changes in the sequence of bases (Fig. 5-1). There are internal DNA repair mechanisms that detect and

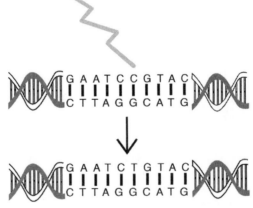

FIGURE 5-1. Mutations can be caused by external factors.

repair mismatched bases, but these can sometimes end up changing a base to match the mutated complementary partner. There is also a low level of errors created by the DNA replication machinery itself. If just 1 in 1 billion bases is copied incorrectly, then every new cell (and every new person) would have a few brand-new mutations.

Mutations that occur in **somatic** cells (all the cells of the body except the sex cells) affect only that cell and its direct progeny within the body of one individual. Mutations in the sex cells (sperm and eggs, or the germline stem cells that produce the gametes) can be passed on to children and become part of the genetic diversity of the human population. In fact, every human is born with a few *new* mutations and many more occur in their somatic cells throughout their lifetime. Since the human population is constantly expanding, just by chance, many new mutations are constantly being added to its store of genetic diversity.

The vast majority of these mutations have no effect on proteins. Remember that only about 1% of the DNA in the human genome is transcribed into RNA and that much of that is spliced out as introns. For those differences that do occur in exons, some lead to no change in amino acids (silent mutations), while others do change amino acids but do not affect the function of the protein (conservative changes). The rare mutations in the coding regions that change the function of a protein are the ones that typically constitute the different observed alleles, such as blue and brown eyes, and single-gene genetic diseases, such as cystic fibrosis and Huntington disease. In fact, genetic disease could be considered to be evolutionary selection in action against mutations that cause significant damage to essential proteins. Many different mutant alleles may exist for a single gene. These may be thought of as different misspellings of the protein sequence, all of which lead to a nonfunctional protein product, which is essential for some physiologically significant process.

Mutations in non-protein-coding regions of DNA can still have phenotypic effects, if they affect intron splicing or gene expression. A mutation in an intron that causes incorrect splicing can produce a nonfunctional protein product just as surely as a coding mutation. A mutation in a promoter or transcription-factor binding site that causes changes in gene expression may have significant phenotypic effects and be just as important in evolution as those that change protein sequences. In rapidly evolving systems, such as defense against pathogens, which can quickly generate new strains, changes in gene expression may provide a more subtle and flexible response than changes in protein sequence. Genetic medicine has concentrated on single-gene inherited diseases that have extremely obvious phenotypes and are often caused by mutations that result in the complete loss of function for a critical protein. However, even among these extreme situations, there are a number of mutations that have been traced to non-protein-coding regions. It is likely that regulatory mutations will play a larger role in multigene complex disorders.

Not all mutations or variants in gene sequences can be cleanly defined as deleterious or beneficial. In some situations, a gene variant may offer an advantage, whereas in another situation, that same variant may be a weakness. The sickle cell mutation of the β-hemoglobin gene is a perfect example. People with a single copy of this allele (heterozygotes) have substantial resistance to malaria, while people with two copies of this allele (homozygotes) experience damaging and sometimes lethal malformations of their red blood cells. In regions of Africa where the incidence of malaria is common, the protection offered to people with a single copy of the sickle cell allele has balanced the damaging effects on people who receive two copies, so that the trait has been maintained in the population. There are obviously other situations in which the trade-off between beneficial and harmful effects of various alleles is less striking. The important concept to keep in mind is that the advantage or disadvantage of a particular mutation depends on the environment in which the individual is living.

Variant alleles that have no apparent phenotypic effect can have an important survival value under some other set of environmental circumstances. For example, the δ-32 variant of the *CCR5* gene provides substantial resistance to HIV. The range of genetic variations present in the whole human population can be considered a reservoir of potential solutions and adaptations accumulated over millions of years that may become useful in some new environment. Advantages may also occur with a particular combination of alleles from several different genes. Our current understanding of human genetics is not sophisticated enough to detect these types of multi-gene interactions, but discovery of these complex traits is one of the primary objectives of plant breeding.

For the large number of mutations that have no effect on protein sequences, splicing, or expression, variants move randomly thorough the human population, following the fate of the section of chromosome that they reside on. These are often called **neutral mutations,** since they have no direct effect on natural selection. However, mutations that do not effect protein-coding regions or gene expression can still be useful diagnostic markers if they are located near genes with important medical effects.

SINGLE NUCLEOTIDE POLYMORPHISMS

A base change at a specific position on the genome is officially considered to be a **polymorphism** in the population when the frequency of the most common base at that position is <99%. These single-base changes are known as single nucleotide polymorphisms (**SNPs**). Sometimes small insertions and deletions are also called SNPs. SNPs are common in the human population. Between any two people there is an average of one SNP every 1250 bases.

These SNPs are potentially valuable as defined markers to track specific regions of a chromosome and possibly as markers for genetic tests. One of the key objectives of the HGP (both the publicly funded project and the private effort by Celera Genomics) has been to identify a large number of human SNPs throughout the genome. This requires comparing the DNA sequence of the same region from many different people. The Celera Genomics project uses DNA from five different people, whereas the HGP uses a much larger number of DNA donors. However, even in the genomic sequence of a single person, SNPs can be identified between the two homologous chromosomes.

An unlikely consortium of pharmaceutical and computer companies have formed a group called the SNP Consortium to pool their resources and develop a large database of human SNPs in the public domain (snp.cshl.org).

> The SNP Consortium Ltd. is a non-profit foundation organized for the purpose of providing public genomic data. Its mission is to develop up to *300,000* SNPs distributed evenly throughout the human genome and to make the information related to these SNPs available to the public without intellectual property restrictions. The project started in April 1999 and is anticipated to continue until the end of 2001.
>
> *Stein* (SNP Consortium website, July 2002. snp.cshl.org)

The January 2001 data release from the SNP Consortium consisted of *856,666* SNPs, all of which have been submitted to GenBank and anchored to the human genome by "in silico" mapping to the genomic working draft (the Golden Path of the University of California at Santa Cruz; see Chapter 4). Together with the International Human Genome Sequencing Consortium, the International SNP Consortium (2001) published a paper in *Nature* that described 1.42 million SNPs that had been archived in a public map base. These SNPs are available on the Web in the integrated genome maps at the NCBI (www.ncbi.nlm.nih.gov), the Ensembl project of the European Molecular Biology Laboratory (www.ensembl.org), the Golden Path genome viewer (genome.ucsc.edu), and the SNP Consortium Web site at Cold Spring Harbor Laboratory (snp.cshl.org/db/snp/map). This huge collection of SNPs spans the entire genome fairly evenly, so that there is a 98% chance of a SNP located within 5 kilobases of every expressed gene.

OTHER MUTATIONS

Other types of common mutations are not confined to a single nucleotide. There are some highly variable sites in the genome that experience frequent mutations. These sites are known collectively as variable number of tandem repeats (VNTRs). There are several different types of VNTRs, which are categorized by the size of the bit of DNA that repeats. **Microsatellites** have repeating units of 2–9 base pairs, and minisatellites have repeating units of 10–100 bases pairs. These sites are characterized by extremely high heterozygosity in the population and instability of the sequence—mutations occur every few generations. These repeats seem to be subject to an inherent flaw in the DNA replication machinery that "slips" in repeat regions.

These VNTR loci are useful as identity markers, such as in the forensic DNA testing that is discussed so frequently in the news and on television crime shows. However, they have limited utility in genomic medicine, since they are so extremely variable, they cannot be used as markers to reliably track other genes. These repeat regions are usually located in noncoding DNA; however, there are a few genes that contain VNTRs that tend to create genetic instability. Fragile X syndrome is an example of a three-base repeat (GCC) in the *FMR1* gene (located on the X chromosome), which has a tendency to expand during meiosis from the normal 20–50 copies up to 200 or more copies. Proteins manufactured from the mutant alleles show loss of

RNA-binding activity, which leads to mental retardation. Other diseases that have been attributed to similar trinucleotide repeats include Huntington disease, myotonic dystrophy, and spinal and bulbar muscular atrophy.

LINKAGE

A mutant allele does not always remain on the same intact chromosome. In addition to the segregation of chromosomes when cells divide, parts of chromosomes crossover between homologous **sister chromatids** in a process known as **recombination** (see Chapter 1). The result of this process is that the set of alleles on a single chromosome gets mixed up over many generations. However, the frequency of recombination between alleles of two genes is a function of the distance between those genes on the chromosome. Therefore, two genes that are located close together on a chromosome will have alleles that are linked, since they will rarely be separated (Fig. 5-2). Thus one allele can be used as a marker for another. This is particularly relevant to genetic medicine, since it is becoming quite easy to find SNP markers linked to genetic diseases (or risk factors) without the (usual) lengthy process of identifying the precise mutation that is responsible for the disease.

More than one cross-over event typically occurs between each pair of chromosomes in each meiosis. However, there seems to be a limit to how close together two recombinations can occur in one meiosis. Therefore, a pair of flanking markers on either side of a target gene can be used to track an important allele with even greater certainty than a single-linked marker (if markers on both sides are present, the chance of a double recombination that removes the central allele is extremely low).

There is a significant amount of basic medical genetics required in order to find the gene or genes responsible for a heritable disease. The starting point for associating a heritable

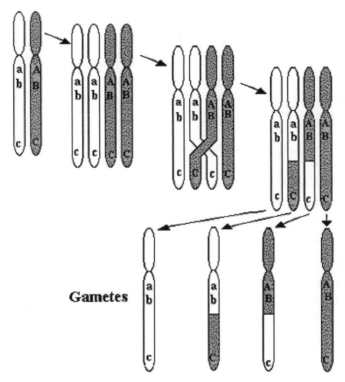

Gametes

FIGURE 5-2. Crossing-over and recombination during meiosis.

disease with one or more genes is a **linkage analysis.** This is usually done by collecting DNA samples from members of families that have multiple instances of the disease—from both affected and unaffected individuals. Then these samples are screened with markers that span all regions of all chromosomes, looking for linkage. Linkage is simply defined as markers that occur in affected individuals more frequently than in unaffected individuals. These markers can be anything that defines a particular allele of a gene—a phenotype, an enzyme activity, a protein of a specific size and chemical identity, or a specific DNA sequence. This work generally requires large families with multiple generations of affected individuals to rule out chance associations and to find markers that are linked tightly to the disease allele (located nearby on the chromosome). It is often desirable to perform linkage analysis on several different families to gain greater statistical power and to confirm results, but it can be difficult to be sure that a phenotypically similar disease in different families is caused by the same genetic defect.

MULTIGENE DISEASES

Many of the most common diseases have been shown to have a significant heritable component yet cannot be traced to alleles of a single gene. These complex diseases, such as asthma, heart disease, and cancer are the result of interactions between many genes (or perhaps just a few). Each of these genes can be considered to be a risk factor for the disease—each contributes, but no single one is either necessary or sufficient to cause disease on its own, or at least not in a significant fraction of patients. However, it is possible to use SNP markers to scan the genomes of families that show inheritance of the disease, or matched groups of people with and without the disease, to discover markers that are linked to these risk factors. In fact, these markers can be used to predict disease susceptibility without ever discovering the identity or function of the culprit genes.

Some complex (i.e., multigene) diseases may actually be caused by any of several different genes scattered across different chromosomes. So it is not that the disease really has multiple causes, but rather that there are multiple diseases that show similar symptoms. SNP linkage analysis can distinguish among these various genetic factors and lead to more precise diagnoses than classical molecular genetic approaches.

GENETIC TESTING

Once SNPs associated with increased disease risk are documented, it will then be possible to screen ordinary patients for these markers. The key limiting factor for the adoption of this technology into routine medical practice is the cost of performing a test for multiple SNPs in a single patient's DNA sample. There are a number of promising technologies currently under intensive development, particularly using DNA microarray approaches. It is also not yet clear how these tests will be integrated into the health-care system. Is it sensible to make a single massive SNP array that can be used to profile each newborn and provide a readout of all disease risks? Or would it make more sense to create specific panels of SNPs that reveal information about susceptibilities to a particular disease and then allow patients and physicians to use the tests when they have a concern about that disease? The presence of a SNP marker that is statistically linked to an increased risk for some disease is not enough information to make important medical decisions, but it does provide a justification to pursue more thorough genetic testing as well as other types of diagnostic procedures.

The frequency of a disease-causing allele (and any markers linked to it) may vary in different human populations. Differences in gene frequencies may be the result of **founder effects,** genetic drift, or some form of natural selection (Fig. 5-3). There are a number of genetic diseases that have well-defined frequencies in particular at-risk populations, such as sickle cell anemia among Africans and Tay-Sachs among European Jews. To use a particular SNP as a genetic marker, it is necessary to establish

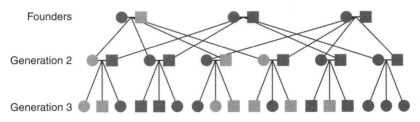

FIGURE 5-3. Gene frequencies change over generations.

the frequencies of the various alleles in different populations. The National Human Genome Research Institute (NHGRI) has established a set of 450 DNA samples that are representative of the genetic diversity found in the U.S. population. These samples can be used to measure the frequencies of SNP polymorphisms and may be useful in establishing standards for genetic tests to validate markers against a common standard.

Every allele of every gene has its own independent evolutionary history (and future) and exists at different frequencies in each subpopulation. Against this background of constant variation and independently evolving alleles, it is important to keep in mind that there is no "correct" sequence. The most common allele for each gene is different in different populations, and it is subject to change over time. Every person has a unique combination of alleles of all of the human genes plus variation throughout all of the noncoding sequence.

Linkage is common and powerful in the human population, particularly in genetically isolated subpopulations, so that a group of alleles for neighboring genes on a segment of a chromosome are often inherited together. Such a combination of linked alleles is known as a **haplotype.** When a new mutation occurs in a singe individual and it is passed down to his or her descendants, it does not move on its own but is carried on a specific a chromosome. Recombination events near the mutant gene are rare, so specific alleles for neighboring genes on that chromosome will remain linked to the mutant gene. Every mutation can be traced back to a single founder chromosome, and it is more likely to be linked to the alleles of nearby genes that were present on that original chromosome than to alternative alleles. The more ancient the origin of the mutation, the less this original linkage can be detected. In theory, over an infinite amount of time, with random breeding of all humans, this linkage would break down. But over periods of hundreds to tens of thousands of years, a state of **linkage disequilibrium** is maintained. So even though every person has a unique combination of alleles of all genes, these alleles are not inherited on a completely random basis—they come in bunches, i.e., haplotypes.

Lander and co-workers at the MIT Genome Center have studied the linkage between SNP markers in human populations (Reich et al., 2001), and they have come up with some surprising results. In a population of pure Northern European descent, most SNPs that were less than 60 kilobases apart showed significant linkage, and half of all SNPs 80 kilobases apart were linked. They also found great variation in the size of linked blocks of DNA across the genome. For example. linkage was detected for 155 kilobases around the Wiskott-Aldrich syndrome-like (WASL) gene, but linkage was noted for <6 kilobases around the protein C inhibitor (PCI) gene. It seems that the genome is composed of a mosaic of recombination hot spots and cold spots, or that each locus on the genome has an intrinsic level of recombination that it supports.

Linkage disequilibrium can be observed in any genetically isolated subpopulation and to a lesser extent throughout the entire human population. Subpopulations that were founded by a small number of individuals, or that have passed through a **bottleneck** that greatly reduced their numbers, have more linkage disequilibrium than larger more thoroughly mixed populations. Thus a medically important allele can more easily be detected by its linkage to other markers within such a subpopulation. Alternately, the predictive value of a specific DNA marker that is linked to a disease allele in one population could be quite different in another population.

Haplotypes could be used to greatly simplify the potential task of genetic testing. A few common haplotypes may be associated with an increased risk for various diseases in specific populations. These haplotypes could be detected with just a few SNP markers. Then genetic disease risks could be evaluated in the context of hundreds of common haplotypes rather than considering all of the individual interactions of tens of thousands of genes, each with its own unique distribution among various subpopulations. Technologies for haplotype testing are currently under development by many biotechnology and genomics companies.

There are other clinical manifestations of human genetic variation. In fact, *all disease has a genetic component, and all therapies should take genetic variation into consideration* (perhaps this can become the motto for the new era of genomic medicine). The physician should be aware of the genetic components of susceptibility vs. resistance to various pathogens, variations in disease severity or symptoms, reactions to drugs (**pharmacogenomics** see Chapter 9), and the variable disease course and prognosis that emerge from the synthesis of all of these factors.

RESEARCH USES OF SNP MARKERS

In addition to the direct medical uses of SNPs in genetic testing (see Chapter 6), these markers also have a valuable role in biomedical research. There are approximately 5000 human diseases known to be caused by the malfunction of a single gene, but so far only a few hundred have been fully characterized as to the normal and mutant gene sequences (see Online Mendelian Inheritance in Man: www.ncbi.nlm.nih.gov/omim). The remaining diseases are extremely rare, making it difficult to find sufficient numbers of families with multiple instances of the disease so that a classic linkage analysis can be conducted. However, with large numbers of SNP markers available, it becomes possible to produce much more precise genomewide screens for common mutations among small populations of affected people, even if they are not related. Common SNP markers, together with a complete human genome sequence, allow researchers to take a positional candidate approach.

Chromosome regions that seem to be common among people with a particular disease can be investigated gene by gene to look for a common mutation with a function related to the disease phenotype.

Pharmaceutical companies are interested in common diseases with multiple risk-factor genes, such as heart disease and cancer. Once again, populations of people who have a disease can be compared to each other and to a group of healthy people to identify SNP markers that are correlated with disease. The chromosome regions near these markers can be examined for genes that are likely candidates for involvement in disease (based on the known function of the protein product or the similarity to proteins of known function). A candidate gene is then sequenced in a small population of diseased and healthy people to see if the diseased people share a common mutation.

Each gene that is identified in this way as contributing to a disease becomes a target for drug-development research. Drugs can be designed to enhance or block transcription of that gene, to interact with the mRNA or protein product of that gene to enhance or block its function, or to interact with the other molecules in the cell with which that protein normally binds. This type of rational drug design promises a great improvement in the efficiency of the drug-development process, leading to improved profits for drug companies and more effective drugs for patients.

ETHNICITY AND GENOME DIVERSITY

The HGP and other related efforts to sequence human DNA have produced some significant findings about human genetic variation that have direct bearing on common notions of ethnic distinctions among groups of people. The most obvious conclusion, as noted earlier, is that the human species as a whole has a much lower level of genetic variation than many other species.

This can be directly attributed to the extremely recent and rapid increase in the human population.

In the human population, genetic heterogeneity is broad, but shallow. There are many sites for genetic variation; but for each locus, a small number of common polymorphisms explain the bulk of the heterozygosity. This distribution indicates that most of the SNPs in the human population are the result of ancient mutational events that created new alleles that have become broadly distributed throughout the population. Since the human population has grown quickly, most of the mutations that it carries date back to a time when there were far fewer humans. In other words, two individuals who share a variant allele have a single common ancestor who was the source of that mutation, even if those two individuals are members of different modern subpopulations. In general, the higher the frequency of a SNP allele in the entire human population, the older the mutation that produced it. Most of the SNPs present at high frequencies in the population were present before humans expanded out of Africa; therefore, it is unlikely that a particular SNP marker will uniquely distinguish any particular ethnic group.

> The gene pool in Africa contains more variation than elsewhere, and the genetic variation found outside of Africa represents only a subset of that found within the African continent. From a genetic perspective, all humans are therefore Africans, either residing in Africa or in recent exile.
>
> *Pääbo (2001:1219)*

The total genetic diversity among the members of any ethnic group is much greater than the diversity between different groups. In fact, 85% of the total human genetic variation is present within any ethnic group. In other words: "It is often the case that two persons from the same part of the world who look superficially alike are less related to each other [in terms of total DNA similarity] than they are to persons from other parts of the world who may look very different" Pääbo (2001:1219) Despite all of the data that show that human races do not exist in any meaningful genetic context, we can see characteristic differences between various ethnic groups—both in the obvious traits (e.g., skin, hair, and eye color and body shape) and in the subtle traits (e.g., the prevalence of hereditary diseases such as Tay-Sachs and sickle cell anemia). These group characteristics can be traced to several factors that influence the genetic makeup of isolated populations over relatively short periods of time. The ancestors of each modern human sub-population were a random selection from a diverse gene pool. By chance, the founders of each sub-group had a higher frequency some variant alleles than the rest of the human population, but overall, they shared no more genetic similarities with each other than with other humans. Over time, selection has strong effects on certain traits within a sub-population, but these common traits represent a tiny fraction of the total genetic diversity in humans.

Certain environmental conditions can produce strong selective pressures that, in small populations, can effect substantial changes in just a few generations. For example, strong sunlight creates a great advantage for genes that darken the skin to prevent burning, whereas weak sunlight favors light skin to enhance vitamin D synthesis. It is also possible in small populations for sexual selection (the favoring by one sex for mates who have a certain trait) to have a strong effect over a relatively short period of time. It is possible to use SNP data to identify chromosomal regions with unusually low levels of diversity in a subpopulation. These are likely to be regions that contain alleles that are currently undergoing selection for favorable mutations—a favorable allele that is increasing in frequency in the population will "carry along" linked markers, leading to a decrease in heterozygosity in that region.

While genome sequence data do not support notions of genetically distinct human races, the data do contain historical information of tremendous value. The pattern of distribution of specific polymorphisms (mutations) can indicate the movements of people over thousands of years. Ancient alleles are widely distributed; more recent mutations have smaller circles of distribution (groups of people descended from a common ancestor). The patterns of sequence variation seen today among various groups of people can be used together with historical, linguistic, and archaeological data to reconstruct their social and genetic histories (Ostrer, 2001).

SOCIAL IMPLICATIONS OF GENETIC DIVERSITY DATA

The human genome is more than a set of instructions for the construction and maintenance of a human being; it is also a historical document. In our genomes, each one of us contains a complete genealogical record of our ancestors, going all the way back to the origins of life. All of the evolutionary selections and the random factors that allowed for the reproductive success of our ancestors are recorded in the DNA sequence of each of our genomes, but this information can be fully interpreted only in the context of comparisons to the genomes of many other human beings and other species. This concept was well expressed by Australian Aboriginal poet Oogeroo Noonuccal (1992):

> Let no one say the past is dead.
> The past is all about us and within
> Haunted by tribal memories, I know
> This little now, this accidental present
> Is not the all of me, whose long making
> Is so much of the past.
> Let no one tell me the past is wholly gone
> Now is so small a part of time, so small a part
> Of all the race years that have moulded me.

The HGP is going to explore this racial history by sequencing DNA from many different people and cataloging the frequencies of polymorphisms in different populations. These data can be read forward to predict many genetic factors related to disease, reactions to drugs, etc. It can also be read backward to deduce the history of human migrations, the isolation and/or intermarriage of different groups, and a great deal more ethnological and anthropological data.

It is also important to keep in mind the limits of genetic factors in determining human individuality. While genes do contribute to the expression of complex characteristics such as intelligence and personality, the complexity of human development must not be oversimplified under the misguided notion of genetic determinism. Just as identical twins have unique personalities, the roles of environmental influences and personal experiences are extremely important in all of the aspects of the development of a person. No one is simply a matrix of interacting genes, and genetic explanations of human behavior are likely to lead to serious errors. In particular, genetic aspects of human behavior and psychiatric illness have been frequently overstated far beyond what the data actually support. It is quite important that medical genetics avoid the loss of credibility that comes with overpromising, so that the genetic tests that truly do offer useful diagnostic and therapeutic value are not equally tainted.

REFERENCES

National Research Council, Committee on Human Genome Diversity. Evaluating human genetic diversity. Washington, DC: National Academy Press, 1997.

International SNP Map Working Group. A map of the human genome sequence variation containing 1.42 million single nucleotide polymorphisms. Nature 2001;409:928–933.

Ostrer H. A genetic profile of contemporary Jewish populations. Nat Rev Genet 2001;2:891–898.

Pääbo S. Genomics and society. The human genome and our view of ourselves. Science 2001;291:1219–1220.

Reich DE, Cargill M, Lander ES, et al. Linkage disequilibrium in the human genome. Nature 2001;411:199–204.

Stein L. SNP Consortium website; http://snp.cshl.org/ 7/17/2002.

Walker K. [Oogeroo Noonuccal]. The past. *In* The dawn is at hand: Selected poems. London: Marion Boyars, 1992: 144p.

GENETIC TESTING FOR THE PRACTITIONER

HARRY OSTRER

Genetic testing is a fairly recent method available to the physician for diagnosing disease and for identifying those at risk for developing a disease or for having a child affected with a disease. Genetic testing has gained wide acceptance in clinical practice, and many genetic-testing laboratories now exist worldwide.

Genetic testing is frequently viewed as different from other kinds of laboratory procedures, because for some diseases genetic testing can identify those who, although currently well, may become ill in the future. Owing to the transmissible nature of genetic information, the identification of a disease-associated mutation has implications not only for the people who are tested, but also for their family members. For these reasons, understanding the implications and limitations of genetic testing are important for both the practitioner and the patient. This chapter provides an overview of genetic testing for the practitioner, including the clinical situations in which it is used, the conceptual basis for the various methods, and the significance of the results.

Essentials of Medical Genomics, Edited by Stuart M. Brown.
ISBN 0-471-21003-X. Copyright © 2003 by Wiley-Liss, Inc.

CLINICAL APPLICATIONS OF GENETIC TESTING

A common question is whether a patient has a certain disease for which there is a genetic basis. Often among the 10,000 conditions for which a genetic component has been identified, the diagnosis can be made on the basis of personal and family history, physical examination, and conventional laboratory tests. A useful reference for identifying these genetic conditions is the Online Mendelian Inheritance in Man (OMIM; www.ncbi.nlm.nih. gov/omim). This catalog is updated regularly and can be searched using multiple terms. The entries provide information about the clinical signs as well as the genetic basis for the condition, if known, including mutations that have been found to cause the condition. To determine whether genetic testing is available for a given condition and to find a laboratory, a useful link is GeneTests (www.genetests.org). The entries in this free online catalog indicate the test menus and contact information for the laboratories and note whether the test is provided on a routine or a research basis.

The clinician is likely to encounter many situations in which a genetic test may be useful. Sometimes genetic testing is required for making a diagnosis, when it cannot be made by clinical criteria alone. The Fragile X syndrome is the most common genetic form of mental retardation. Although the diagnosis may be suggested by the presence of the characteristic signs—large ears, protruding chin, and large testes—the only way in which the diagnosis can be made is by genetic testing. Although the clinical characteristics of the various forms of spinocerebellar ataxia overlap considerably, they can be readily distinguished by their specific mutations. Patients with atypical forms of certain diseases may have a negative gold standard but positive genetic test. For example, most patients with cystic fibrosis are diagnosed by a sweat chloride test. However, some individuals who have pulmonary disease suggesting this condition have a normal sweat chloride test. For these patients, the diagnosis is made on the basis of finding mutations in both copies of their cystic fibrosis transmembrane conductance regulator (CFTR) genes.

Sometimes a patient has not shown any signs of a particular condition, but because of his or her family history, the patient may want to know about his or her risk of developing disease. This is common, for example, for people who have lost a parent to Huntington disease (a progressive neurodegenerative disease) and for women whose mother and/or sister may have died from breast or ovarian cancer, suggesting a heritable risk. For these individuals, a positive genetic test result indicates an increased, although not necessarily absolute, risk for developing the disease.

Genetic testing can be used for assessing reproductive risks by testing the parents for carrier status and by testing the fetus. Individuals with a positive family history of genetic disease (usually autosomal recessive or X-linked) or who come from ethnic groups with an increased prevalence of particular autosomal recessive diseases are candidates for carrier screening. Currently, carrier screening for cystic fibrosis is recommended in the United States. For people of Mediterranean, African, and South Asian ancestry, hemoglobinopathy screening is recommended. Individuals whose ancestors were Ashkenazi Jews may be screened for Tay-Sachs, Canavan disease, cystic fibrosis, Gaucher disease, Bloom syndrome, Fanconi anemia, Niemann-Pick disease, and/or familial dysautonomia.

If an individual is a carrier for a certain condition, he or she may choose not to have children with a person who is also a carrier for the same condition. If both the man and the woman are carriers, they may choose to have prenatal diagnosis to determine whether their fetus is affected by the condition. Fetal testing can be performed at 10–11 weeks of gestation, using the chorionic villus sampling procedure, by which bit of placenta (made of fetal tissue) is obtained under ultrasound guidance. Alternatively, an amniocentesis can be performed at 15–18 weeks of pregnancy to obtain fetal cells from the amniotic fluid. The fetal DNA obtained from these tests is examined for the disease in question.

Not all genetic testing involves looking for heritable mutations. Sometimes it is used to look for genetic alterations that are confined to a specific population of cells. Such alterations may cause certain cells to become cancerous or may cause cancerous cells to progress to a more aggressive form. Genetic testing can be used to identify chromosomal translocations between two nonhomologous chromosomal segments. For example, a translocation between chromosomes 1 and 19 found in cancerous cells of a leukemia patient indicates the acute promyelocytic form of the disease; whereas a translocation between chromosomes 9 and 22 is characteristic of the chronic myelogenous form.

METHODS OF GENETIC TESTING

DIRECT MUTATION TESTING

The method of genetic testing has to be geared to the condition that is being detected. Some heritable diseases are caused by a single or limited number of well characterized mutations. In small population groups, a mutation may have achieved a carrier frequency of 1% or greater, owing to **founder effects** (see Chapter 5). For these conditions, genetic testing is relatively efficient because it is geared to the detection of specific mutations. Other conditions, however, are produced by a wide variety of different mutations, and screening methods used for them must be able to detect all of the possible sequence variants.

Most genetic testing involves the use of polymerase chain reaction (PCR), whereby many copies of a gene sequence are made from DNA primer molecules that define the ends of the sequence to be amplified (see Chapter 2). If **allele specific primers** are used, success in producing multiple copies of the target fragment can be diagnostic for the presence of a translocation, deletion, or insertion. In other cases, the size of the amplified fragment can be diagnositc for the presence of a deletion or insertion mutation. Single nucleotide polymorphisms (SNPs), involving base-pair changes, can be detected by a variety of techniques, including allele-specific PCR, primer extension, and ligation, or hybridization of a probe directly at the site of the base change (Fig. 6-1). The specific reaction occurs only when there is a perfect nucleotide match between a probe and a target sequence in the patient's DNA. These techniques

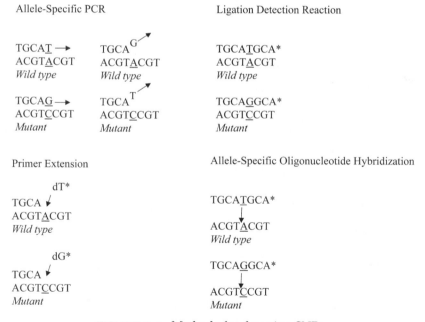

FIGURE 6-1. Methods for detecting SNPs.

can be used to detect both common and uncommon (or wild-type and mutant) alleles at a given genetic locus.

Many of these methods now lend themselves to multiplexing, whereby alleles at several different sites are tested at the same time. In the multiplexing format, PCR can be performed on several different fragments simultaneously, and it is possible to test for mutations at more than one site in each of these fragments. These methods are useful when the mutation sites to be queried are known. A multiplex test may involve many different sites of possible disease causing mutations within a single gene, or may involve tests for mutations in many different genes.

When the sites of mutation are unknown, other methods are used to test for sequence variation in the genome. The gold standard, and most commonly used, method is DNA sequencing. Sequencing can be performed directly on PCR-amplified DNA fragments. This analysis can be quite efficient, as up to 800 base pairs can be analyzed in a single PCR reaction. Other methods take advantage of changes that occur in the physical properties of DNA (as the result of heteroduplex formation by hybridization). when a mismatch occurs between an amplified fragment of DNA from a patient and a reference fragment that contains the normal sequence of that same gene. The commonly used methods include single-stranded conformation polymorphisms (SSCP) and denaturing high-pressure liquid chromatography (dHPLC). Generally, these are deemed screening methods, and any newly detected polymorphism is confirmed by DNA sequencing.

One of the major challenges of sequencing-based strategies is the interpretation of novel DNA sequences. If a novel sequence can be shown to disrupt the coding region of a gene by introducing premature terminators, then the sequence is usually deemed to be a mutation. If such a change alters the amino acid sequence, it may be unclear whether this mutation disrupts protein function and is the cause of the observed disease phenotype. The role of a novel sequence may be clarified by case-control studies in which the specified mutation is shown to be present in a high proportion of affected individuals but a low proportion of normal controls. Linkage studies may reveal that the specific mutation is linked to the transmission of the observed phenotype in a family.

LINKAGE ANALYSIS

In the past, linkage analysis was commonly used as an alternative to direct mutation methods. Linkage was used when the map position, but not a disease-causing gene, was known. Generally, members of three generations in a single family participated in the study so researchers could be confident about the assignment of phase for the markers (Fig. 6-2). Markers that flanked the region of interest were chosen to account for

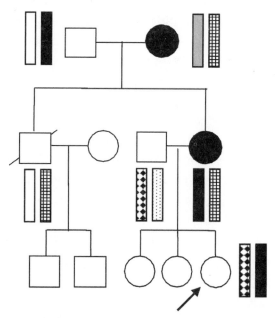

FIGURE 6-2. Linkage analysis can detect individuals who are carrying disease-conferring mutations.

recombination that could limit the accuracy of prediction. Linkage analysis has fallen by the wayside in clinical applications because the genetic basis for many diseases has been identified. However, linkage remains a powerful research tool for identifying new genes that may cause specific conditions that are transmitted through particular families.

ADEQUACY OF GENETIC TESTING

Genetic tests should be designed that are both sensitive and specific—that is, a high proportion of cases should be detected by a direct or indirect mutation test in a gene (sensitivity), and individuals without the mutation should not be erroneously diagnosed by the test (false positive results). Despite the best of intentions, this is not a foregone conclusion. The available methods can preferentially determine sequence alterations of a certain type but may miss others. Some conditions are caused by a variety of different mutations in a single gene or by mutations in several genes and thus may be missed. Genetic testing needs to be carried out with appropriate controls so that positive and negative results can be analyzed.

Even in the case of higly sensitive and accurate tests which detect a specific mutation, disease diagnosis may be problematic. Some genetic diseases have variable levels of penetrance – individuals with the mutation may exhibit different levels of disease symptoms depending on the modifiying effects of other genes, environmental and other factors. For example, patients with the mutant forms of the BRCA1 or BRCA2 gene have elevated risk of breast cancer, but the gentic diagnosis does not provide firm information when, if ever, the disease will occur.

Because of the sophistication required for genetic testing, regulatory measures have been imposed both by state health departments and by the Health Care Finance Administration, specifically under the regulatory authority that was provided by the Clinical Laboratory Improvement Act of 1988 (CLIA 88). This regulation specifies the qualifications for laboratory directors, supervisors, and technicians and the requirements for standard operating procedures for testing, reporting, quality assurance, and quality control. Many jurisdictions now require that testing be performed in CLIA-approved laboratories if the results are to be reported to patients.

The Secretary's Advisory Committee for Genetic Testing (2001), appointed by the secretary of Health and Human Services has recommended national standards for genetic testing aimed at improving not only test production but also test validation. Such standards could lead to the large-scale marketing of genetic test kits, much as occurs for other forms of contemporary laboratory testing.

INFORMED CONSENT

Genetic testing is viewed as exceptional compared to other forms of laboratory testing, because it does not vary over time and it provides information about individual's innate heritable predispositions, which are not under the individual's control. In addition, this information has implications not only for the person tested but also for his or her family members. For these reasons, many states have passed laws that require individuals to provide informed consent before a genetic test is performed. These laws require that testing be performed only for the conditions for which consent was provided. Some laws allow for residual DNA samples to be available for research purposes, if the donor remains anonymous.

GENETIC COUNSELING

Patients commonly do not fully understand the significance of genetic testing. It is critical before testing to explain to a patient the reasons for performing the test, the natural history of the condition being tested for (including possibilities for intervention), and the significance of positive and negative results. After testing, it is important to provide additional counseling about the meaning of the test results, whether negative or positive. A positive result does not necessarily ensure that an individual will develop a particular condition. Likewise, a negative result does not mean there is no risk.

Genetic testing for the diagnosis of inherited diseases is a powerful tool for the practitioner. The technology is advancing rapidly, and the number of diseases that can be diagnosed via these methods is growing. For the practitioner, the best approach is to understand the principles on which the tests are based and to apply them on a disease-by-disease basis. Decisions about genetic testing can be facilitated by consultation with an experienced medical geneticist or genetic counselor. Careful application of DNA analysis in the proper setting can improve patient care dramatically.

REFERENCE

Secretary's Advisory Committee on Genetic Testing. Enhancing the oversight of genetic tests: Recommendations of the SACGT. www4.od.nih.gov/oba/sacgt.htm. Accessed July 2001.

SUGGESTED READING

Andrews LB, Fullarton JE, Holtzman NA, Motulsky AG, eds. Assessing genetic risks: Implications for health and social policy. Washington, DC: National Academy Press, 1994.

Buetow KH, Edmonson M, MacDonald R, et al. High-throughput development and characterization of a genomewide collection of gene-based single nucleotide polymorphism markers by chip-based matrix-assisted laser desorption/ionization time-of-flight mass spectrometry. Proc Natl Acad Sci USA 2001;98:581–584.

Grody WW, Cutting GR, Klinger KW, et al. Laboratory standards and guidelines for population-based cystic fibrosis carrier screening. Genet Med 2001;3:149–154.

Hacia JG, Brody LC, Chee MS, et al. Detection of heterozygous mutations in BRCA1 using high density oligonucleotide arrays and two-colour fluorescence analysis. Nature Genet 1996;14:441–447.

Kronn D, Jansen V, Ostrer H. Carrier screening for cystic fibrosis, Gaucher disease, and Tay-Sachs disease in the Ashkenazi Jewish population: The first 1000 cases at New York University Medical Center. Arch Intern Med 1998;158:777–781.

Motulsky AG. If I had a gene test, what would I have and who would I tell? Lancet 1999;354(Suppl 1):SI35–37.

Motulsky AG. Screening for genetic diseases. N Engl J Med 1997;336:1314–1316.

Ostrer H. A genetic profile of contemporary Jewish populations. Nature Rev Genet 2001;2:891–898.

Ostrer H, Hejtmancik JF. Prenatal diagnosis and carrier detection of genetic diseases by analysis of deoxyribonucleic acid. J Pediatr 1988;112:679–687.

Pandolfi PP. Oncogenes and tumor suppressors in the molecular pathogenesis of acute promyelocytic leukemia. Hum Mol Genet 2001;10:769–775.

Saiki R, Scharf S, Faloona F, et al. Enzymatic amplification of B-globin genomic sequences and restriction site analysis for diagnosis of sickle cell anemia. Science 1985;230:1350–1354.

Wiggins S, Whyte P, Huggins M, et al. The psychological consequences of predictive testing for Huntington's disease. Canadian Collaborative Study of Predictive Testing. N Engl J Med 1992;327:1401–1405.

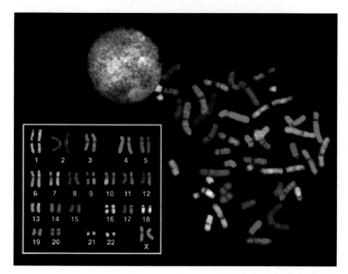

FIGURE 1-1. The human karyotype (SKY image).

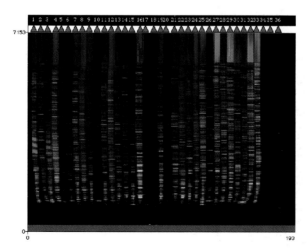

FIGURE 2-10. A fluorescent sequencing gel produced on an automated sequencer. Each lane contains all four bases, differentiated by color.

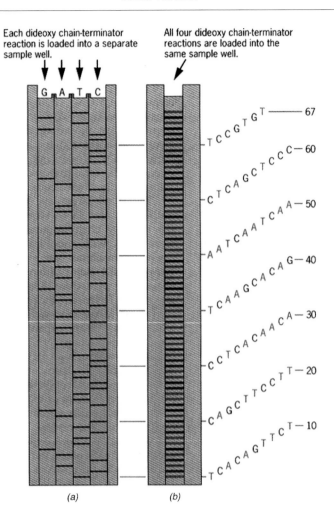

Each dideoxy chain-terminator reaction is loaded into a separate sample well.

All four dideoxy chain-terminator reactions are loaded into the same sample well.

G A T C

$T \text{—} 67$
$TCCGTG$

$C \text{—} 60$
$CTCAGCTCC$

$A \text{—} 50$
$AATCAATCAA$

$G \text{—} 40$
$TCAAGCACA$

$A \text{—} 30$
$CCTCACAAC$

$T \text{—} 20$
$CAGCTTCCT$

$T \text{—} 10$
$TCACAGTTC$

(a)   (b)

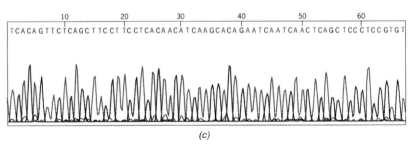

10   20   30   40   50   60

TCACAGTTCTCAGCTTCCT TCCTCACAACATCAAGCACAGAATCAATCAACTCAGCTCCCTCCGTGT

(c)

FIGURE 2-11. ABI fluorescent sequencers allow all four bases to be sequenced in a single gel lane and include automated data collection.

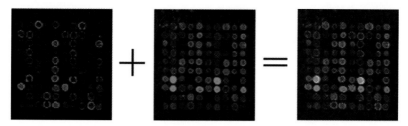

FIGURE 8-2. Two separate fluorescent microarray (with red and green false colors) are combined to show the relative gene expression in the two samples.

FIGURE 8-7. A spotted cDNA array hybridized with a mixture of two probes and two different fluorescent labels visualized as a red–green false-color image.

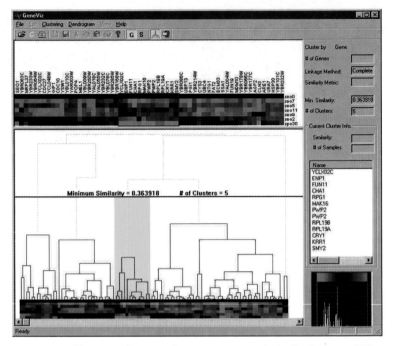

FIGURE 8-8. Clusters of genes that are expressed similarly over different experimental treatments. (Reprinted with permission from Seo and Lee, 2001.)

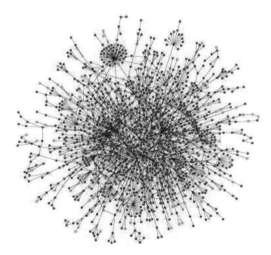

FIGURE 10-2. A map of protein–protein interactions for 1870 yeast proteins. (Reprinted with permission from Jeong et al., 2001.)

GENE THERAPY

JOHN G. HAY

HISTORICAL PERSPECTIVE

The concept of using DNA as a treatment for genetic disease—called *gene therapy*—has its roots in the history of genetics and molecular biology. Studies performed in 1928 by Griffith, a British microbiologist, demonstrated that when heat-killed bacteria are mixed with live bacteria the characteristics of the living bacteria can change. Avery Macleod and MacCarty (1944) showed that DNA was the "transforming principle" responsible for changing harmless bacteria into a pathogenic form. The principal that the addition of DNA can alter the phenotype of an organism was thus established by the middle of the twentieth century.

Subsequent studies established the gene as the unit of inheritance and determined that specific gene mutations could lead to specific inherited diseases. The rapid developments in **recombinant DNA cloning** (or genetic engineering) in the 1970s and 1980s allowed gene replacement within an individual to become a realistic possibility. Researchers began to develop strategies for gene therapy based on recombinant DNA technology. For instance, the gene mutation that is responsible for the

Essentials of Medical Genomics, Edited by Stuart M. Brown.
ISBN 0-471-21003-X. Copyright © 2003 by Wiley-Liss, Inc.

clinical manifestations of cystic fibrosis was identified in 1989. By 1992, the initial clinical trials for using viruses to replace the defective gene in cystic fibrosis patients were already in progress.

Since the 1990s, the early optimism for gene therapy has been replaced with a more temperate appraisal and a greater understanding of the problems that remain to be overcome. However, several recent studies have reaffirmed the hope that gene replacement may have the potential to become a viable therapy for human diseases in the not too distant future.

STRATEGIES OF GENES THERAPY

The most intuitive application for gene therapy is to correct an inherited defect within a single gene that causes a genetic disease. Since the cause of the disease is clear—a mutation within a single gene—the potential therapy is equally apparent—replace the faulty gene with a normal copy. However, most cells in the body have a limited life span and do not replicate, so an introduced gene would have only a temporary effect. For a gene therapy strategy to achieve a permanent cure, stem cells must incorporate the new gene and continually supply progeny cells with the corrected genotype. Introducing an engineered gene into human germ cells (or even into embryos) would lead to the production of genetically engineered humans, creating a permanent change in the human species. In view of the ethical and safety concerns of germ-line transmission of genetic alterations, all studies performed so far have focused on **somatic**-cell gene transfer. The consequence of this approach is that gene transfer is not permanent and the therapy needs to be repeated.

Initial gene-transfer studies focused on diseases caused by single-gene mutations. Some examples are cystic fibrosis, familial hypercholesterolemia, and adenosine deaminase deficiency. Gene therapy research has also examined diseases that result from alterations in many genes, including cancer and cardiac and limb ischemia. The strategy in these applications is to deliver a new therapeutic gene rather than to correct an inherited abnormal gene.

Another major area of interest, is to use genes to make tumors more immunogenic, or to enhance the immune response against a tumor. Genes are also being used to induce an immune response against potential pathogens, in essence using genes as vaccines. Another area of gene therapy involves creating genetically modified viruses to replicate specifically within tumor cells.

DNA ELEMENTS FOR GENE EXPRESSION

Assuming that a gene can be delivered to the nuclei of appropriate cells within target tissues, certain essential DNA elements are required for gene transcription (Fig. 7-1). The first essential element is the coding sequence of the gene. Depending on the size of the gene and the capacity of the vector used to deliver the gene into the cell, either the genomic sequence (including the noncoding **introns**) or the **cDNA** (the protein coding sequence with introns removed) is used. The **promoter** and additional 3′ and 5′ flanking sequences, important for initiation of transcription and translation and RNA processing, are also needed.

The promoter is a stretch of DNA sequence that directs the transcription of the gene. It can limit transcription to a particular

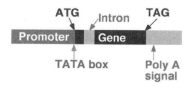

FIGURE 7-1. DNA elements required for gene transcription include a TATA box (for RNA polymerase binding), polyadenine site (responsible for signaling the addition of a polyadenine tail to the transcript), ATG codon (for initiation of translation), and TAG codon (translation stop codon).

tissue, to a particular period of development, or to a particular phase of the cell cycle. The promoter can also modify transcription in response to particular external triggers (e.g., radiation). The promoter provides a docking site, or TATA box, for the RNA polymerase–associated protein complex as well as transcription-factor binding sequences which attract proteins that enhance (transactivators) or repress (repressors) transcription. Ideally, the DNA segment to be transferred should contain a gene with a tightly regulated promoter; however, limitations of current vector systems and the effects of surrounding DNA **enhancer** elements in those vectors make it difficult to achieve specificity of gene expression. Thus a simpler, "always-on" promoter is often used. As gene therapy experiments become more sophisticated, more precisely controlled promoters are likely to be used.

GENE DELIVERY SYSTEMS

The basic challenge of gene therapy is to develop a mechanism by which a new gene is placed in a patient's target cells in a form in which it can be expressed. Thus the introduced gene must enter the nucleus of the target cell, which is accomplished in several steps: The gene is delivered to the cell surface, binds to the cell membrane, crosses into the cell's interior, negotiates the intracellular trafficking pathways to reach the nuclear membrane, and enters the nucleus. Once the gene is in the nucleus, its DNA must be unpacked and released in a transcription-ready form (Fig. 7-2). Of course, each step presents its own obstacles, including the body's protective mechanisms that prevent foreign material, particularly genetic material, from gaining access to a cell's nucleus.

The many delivery systems that have been studied can generally be divided into viral and nonviral methods. In either case, the predominant problem has been introducing enough

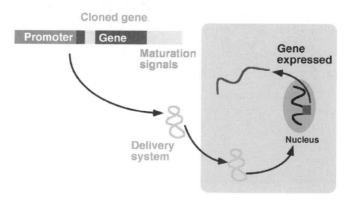

FIGURE 7-2. Introducing a new gene into the cell's nucleus.

DNA into enough cells in the target tissue to achieve therapeutic levels of gene expression. An additional problem has been balancing the level of gene expression against any potential toxicity of the delivery vehicle

NONVIRAL DELIVERY SYSTEMS

NAKED DNA   The simplest delivery system consists of a gene cloned into a plasmid that can be amplified in bacteria, purified, and then administered, as "naked" DNA. Because this method of DNA delivery is highly inefficient at **transfection,** (transporting the gene from the cell surface to the nucleus), levels of gene expression are extremely low. However, even low levels of protein expressed from the introduced gene is sometimes sufficient to induce an immune response within the host. This approach may, therefore, lend itself to using DNA as a vaccine. For instance, if a gene that encodes a viral protein from a hepatitis virus can be transfected and expressed, it may be possible to induce a protective immune response against that virus.

PARTICLE BOMBARDMENT   Bolistic particle-mediated delivery systems (gene guns) are being evaluated as a method for

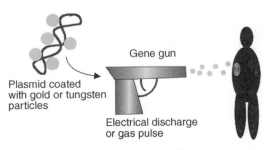

FIGURE 7-3. Particle bombardment.

improving delivery of naked DNA molecules into cells (Fig. 7-3). In these systems, a DNA sequence of any length is coated onto small particles of either gold or tungsten; the particles are then fired at the tissue via a helium pressure device, or gun. The target tissue needs to be exposed, so skin and wounds are good targets. Transgene expression can be observed in both epidermis and dermal compartments after bolistic gene delivery to intact skin. Efficiency of gene delivery is low, but enough protein can be expressed to induce an immune response. Applications that are being investigated for this system include immunization against influenza, viral hepatitis, HIV, and tuberculosis. Research is also focusing on delivery of genes for growth factors to wounds and therapeutic genes to melanomas on the skin surface. Particle bombardment has successfully been used to deliver genes to plants and fish.

LIPOSOME VECTORS   Although lipids with a neutral, negative, or a positive charge can complex with DNA, the positively charged—or cationic lipids—have shown the most promise in facilitating gene delivery. When these lipids complex with DNA, they form small globules known as liposomes. Cationic liposomes can attach to negatively charged DNA and still maintain an overall positive charge, which facilitates attachment to the negatively charged cell surface. A complex of cationic lipids and nucleic acid is referred to as a lipoplex. Two examples of cationic lipids are N-[1-(2,3-dioleoyloxy)propyl]-N,N,N-trimethylammonium chloride (DOTMA) and 3b-[N-(N,N-dimethylaminoethane)carbamoyl] cholesterol (DC-Chol). If a cationic polymer is used in place of a cationic lipid, the DNA conjugate is called a polyplex; two cationic polymers are polyethylenimine and polylysine. Lipopolyplexes are combinations of cationic lipid, cationic polymer, and DNA.

Unfortunately, all these complexes face similar problems. First, the complex must be made, and this process remains empirical and depends on many factors, including charge, relative proportions of DNA and lipid/polymer, and ionic strength of the solution. The final structure is often unclear, but one favored structure for lipoplexes is for the DNA to be intercalated within two lipid bilayers (Fig. 7-4).

Because liposomes are charged particles, it can be difficult for then to traverse the tissue matrix and reach the target cells. Once at the target cell, the liposome must fuse with the cell

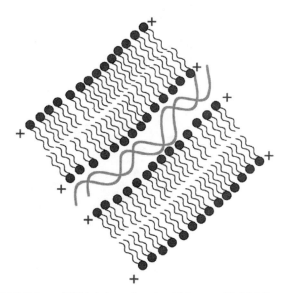

FIGURE 7-4. DNA intercalated within two lipid bilayers.

surface or cross the surface membrane to enter the cell Endocytosis is probably the major pathway that DNA in liposomes uses to enter the cell (Fig. 7-5). Once the particle has entered the cell, the next step is to free the DNA from the complex. Most liposome formulations include a fusigenic co-lipid, which facilitates the release of the DNA. Once free, the transfected DNA needs to reach the nucleus, cross the nuclear membrane, and be disassembled from any remaining carrier before transcription can begin. Nuclear localization sequences on the transfected DNA segment may help the DNA cross into the nucleus.

Liposomes have shown considerable promise **in vitro** (i.e., cell culture), but many obstacles must be overcome before we can achieve success with **in vivo** applications. Serum can affect the stability of the liposomes, rapidly eliminating them from the bloodstream. Furthermore, the particle may trigger the

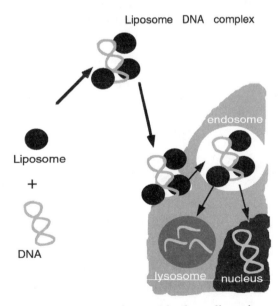

FIGURE 7-5. Liposomes must fuse with the cell surface or cross the membrane. Once in the cell, the complex must break out of the endosome and then translocate across the nuclear membrane.

blood-clotting mechanism, with toxic effects. Additional problems associated with the use of **in vivo** liposome vectors are their low efficiency of nuclear transport and the instability of the DNA in the cytoplasm.

VIRAL DELIVERY SYSTEMS

Over many millions of years, viruses have evolved mechanisms to overcome the obstacles blocking delivery of genetic material into the cell's nucleus. Thus viruses make attractive vectors for gene therapy. Although all viruses share some similarities in the methods used to transfer their genetic material, it is the distinct properties of specific viruses that dictate the choice of vector suitable for a particular gene therapy application.

Viruses usually infect cells by targeting a surface receptor that is used by the cell for other functions. This receptor targeting improves internalization of the virus into endosomes within the cell. Viruses have also developed mechanisms for escaping from the endosome in the cytoplasm and avoiding lysosomal degradation. For instance, the adenovirus achieves this through a viral structural protein (penton) that causes endosomal rupture in a low pH environment. Viral DNA commonly contains nuclear localization signals to facilitate transport to the nucleus. Some viruses, such as retroviruses, are able to integrate their genes into the chromosomes of host cells, potentially allowing the expression of viral genes for the life of both the host cell and that cell's progeny. Some viruses, however, express their genes from episomes, which do not persist in the host cell, and thus these organisms cannot provide a vehicle for permanent gene therapy.

RETROVIRUSES   The retroviruses are RNA- viruses, and the one most commonly used in gene therapy is the murine leukemia virus. The main benefit of using a retrovirus as a vector is its ability to use reverse transcription and **recombination** to integrate a transferred gene permanently into the host cell chromosome, where it will remain through subsequent cellular divisions, or until cell death.

Unfortunately, this enormous benefit is associated with serious problems. In particular, there is no control over the site of integration in the host cell's genome, which can lead to disruption of a gene at the site of insertion. Furthermore, the transferred promoter/enhancer sequences can influence the expression of the genes surrounding the site of insertion. Disruption of the expression of a tumor-suppressor gene and enhanced expression of an oncogene are clearly grave situations that could lead to the development of a malignancy. Although the viruses used for gene therapy are not able to replicate on their own, the creation of a dangerous replication-competent retrovirus during manufacture by recombination of the vector with packaging cell components is also a concern. Another drawback is that the retrovirus, when produced in a murine cell line, is not stable in human blood and is subject to complement-mediated inactivation, which limits delivery options.

In addition, the target cell must be replicating for integration to occur. This requirement impedes application of this vector to many tissues that have slow cellular turn over, such as the respiratory epithelium. However, this may also be a benefit when specifically targeting growing cancer cells. Despite persistence of the integrated DNA within the genome of the cell, there is no guarantee of the persistence its expression. Epigenetic events, in particular DNA methylation, can silence the transferred gene.

The steps taken to produce a retroviral vector are shown in Figure 7-6. The cDNA of the therapeutic gene is cloned into a gutted retroviral genome, which has been stripped of the essential *gag, env,* and *pol* genes between the inverted terminal repeats. This modified genome is then transfected into a

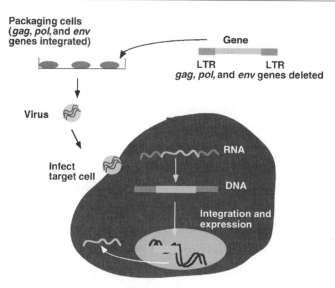

Packaging cells
(*gag, pol,* and *env*
genes integrated)

Gene

LTR   LTR
gag, pol, and *env* genes deleted

Virus

Infect
target cell

RNA

DNA

Integration and
expression

FIGURE 7-6. Creating a retroviral vector.

producer cell line, which is a line that can be grown in culture containing the essential missing retroviral genes. The resulting virus is infectious but cannot replicate. Once the engineered virus reaches the nucleus of the target cells, viral RNA is reverse transcribed into DNA, and the modified viral DNA is integrated into the host genome. The cloned therapatic gene can then be expressed in the host cell.

Retroviruses do not live long within the circulation, so researchers have developed several different approaches for gene delivery. Host cells, particularly lymphocytes and tumor cells, have been removed from the body, infected *ex vivo* and then readministered to the host. In another method, producer cells themselves are administered directly into certain tumors, such as gliomas. The infected producer cells die within the tumor, releasing the altered virus to infect preferentially the replicating glioma cells.

ADENOVIRUS   Because the biology of adenovirus provides many properties that are well suited for gene therapy, this virus is currently the most commonly used vector. Adenovirus is a DNA virus with a 36-kilobase double-stranded DNA genome. It is relatively easy to produce high titers of infectious virus, and it is possible to insert reasonably large genes. The virus efficiently infects many cell types, including nondividing cells, and is more stable in the circulation than is retrovirus.

Adenovirus infects the target cell by two interactions on the cell surface. The adenoviral fiber proteins, which protrude like spikes from the surface of the viral particle, interact with the high-affinity adenoviral–coxsackie receptor (CAR) on the cell surface (Fig. 7-7). CARs are found on the surface of most cell types, although some tissues have more than others. However, after intravenous administration, preferential deposition of adenovirus from the bloodstream in hepatic and pulmonary capillary beds appear more important than levels of CAR expression. The second interaction on the cell surface is between the viral

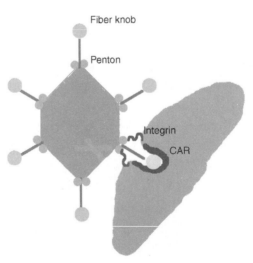

FIGURE 7-7. The adenovirus infects the target cell through two interactions on the cell surface.

penton proteins, located at the base of the fiber shafts, and the integrins, on the cell surface; this interaction triggers viral internalization into into the cell by endocytosis.

The viral capsid is highly efficient at escaping from the endosome, which ruptures in low pH environments. The path to the nucleus is also fast and efficient; utilizing microtubules to deliver the capsid to the nuclear surface. The viral DNA is then injected through the nuclear membrane into the nuclear interior and is ready for transcription to begin. The adenovirus DNA does not integrate into the genome but remains episomal. Infection is, therefore, transient, necessitating repeated administration for persistent gene expression. This is a problem, since the virus induces a potent cell-mediated immune response, and neutralizing antibodies may limit the success of repeated doses.

Most adenoviral gene therapy vectors are made safe by removing a gene essential for viral replication. However, as will be discussed later in this chapter, some experiments with partially replication-competent virus have been attempted. The steps for constructing a replication-deficient adenoviral vector are shown in Figure 7-8. The *E1a* regions of the viral genome, which is essential for viral replication, is deleted and replaced with the cDNA of the therapeutic gene. The viral genome is then transfected into a producer cell line that has the viral *E1a* gene integrated in its chromosomal DNA, thereby providing the essential functions lacking in the modified viral genome. Infectious but replication-deficient virus for gene therapy applications is purified from the cell line.

An inflammatory response or an immune response against the viral proteins is a problem with these first-generation viruses. Several modifications have thus been introduced, for example, the deletion of other viral genes such as *E4* or *E2* that encode proteins that can induce a an inflammatory immune response. Specialized producer cell lines that express the

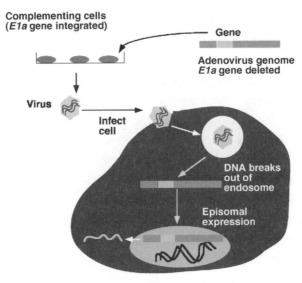

FIGURE 7-8. Construction of an adenoviral vector.

additional deleted genes (which are often toxic) are required to package these second-generation viruses. This line of research has culminated in a viral genome that is virtually devoid of all viral genes— a "gutless" adenovirus— that depends on a helper virus for production.

In recent years there has been great interest in using conditionally replicating adenoviral vectors. The genes for replication remain essentially intact in these vectors, but the virus is designed to replicate only in certain cell types. These strategies are discussed later in this chapter.

ADENO-ASSOCIATED VIRUS   Adeno-associated viruses (AAV) are defective parvoviruses that were first noticed as contaminants in adenoviral laboratory stocks. AAV is a single-stranded DNA virus that is dependent on a helper virus for the completion of its life cycle. Although the role of the helper virus is not completely clear, it appears to play role in making a cellular environment conducive to AAV replication. AAV have several properties that are favorable for gene therapy vectors: They infect both dividing and nondividing cells with reasonably high efficiency, do not cause human disease, and do not induce an inflammatory response. Although the human host can generate neutralizing antibodies, the immune response against AAV is limited.

AAV binds to heparin sulfate proteoglycans on the cell surface, and both integrin $\alpha v\beta 5$ and human fibroblast growth factor 1 can act as co-receptors. Wild-type AAV, in the absence of the helper virus, integrates into the host cell genome and, in a site-specific way, to chromosome 19. However, integration depends on the viral *rep* gene, and when the viral genome is modified to carry a transgene, integration is rarely seen. The capacity of AAV to accept a transgene is limited to approximately 4.5 kilobases. AAV has been most actively investigated for clinical roles in cystic fibrosis, hemophilia, and ophthalmic disease.

HERPES VIRUS Herpes simplex virus is a double-stranded DNA virus with a 152-kilobase genome. Its large size is conducive to the insertion of large transgenes and even multiple transgenes. The virus is able to undergo a persistent, nonintegrated latent infection of neuronal cells, which occurs without expression of viral lytic proteins. However, a neuronal-specific promoter is active during latency and thus has the potential for long-term expression of a transgene. The neural tropism of herpes simplex has led to its choice as a vector for gene therapy of neurodegenerative diseases. The virus can, however, infect a wide variety of other cell types in both the dividing and the resting state. The use of this vector for muscular dystrophy has been considered, because herpes may be able to transfer the large (14 kilobase) dystrophin gene. Applications for cancer therapy, both as a vector for a toxic gene in a replication-deficient virus and as a conditionally replicative lytic virus are being investigated.

The herpes simplex virus has a complex structure, making it difficult to target. A protein cover and an envelope of glycoproteins surround the viral capsid. The virus attaches to heparin sulfate and glycosaminoglycans on the cell surface by an interaction with the glycoproteins. Instead of crossing the cell membrane, the virus fuses with it. Several receptors have been identified as playing a potential role in viral internalization.

VACCINIA VIRUS   Vaccinia is a double-stranded DNA virus of the pox family with a 200-kilobase genome; it is unusual in that replication and transcription occur in the cytosol, so integration in the host cell nucleus does not occur. The virus has a broad tropism and allows the insertion of large transgenes. Vaccinia induces a vigorous immune response and, although comparatively safe, can be fatal in immunosuppressed individuals. This vector is being investigated as an agent for delivering cytokine genes to tumors, including bladder cancer and melanoma.

ALPHAVIRUS   Alphaviruses are RNA viruses that can infect a broad host range of dividing and nondividing cells. Gene expression is at high levels but transient, as no integration occurs. The Sindbis virus is the alphaviruses that has received the most attention as a gene therapy vector. Alphavirus are being investigated for gene transfer to tumors and for vaccine development.

LENTIVIRUS   Lentiviral vectors are derived from HIV but, unlike other retroviruses, are able to transduce nondividing cells. Lentivirus vectors can deliver up to 8 kilobases of transgenic DNA, which becomes integrated and thus is expressed long term in the host cell without inducing a host immune response. Lentivirus vectors carrying therapeutic genes are being evaluated for β-thalassemia and Parkinson disease.

HYBRIDS From the preceding discussion it can be inferred that individual viruses have both highly desirable properties for use as gene-transfer vectors but also some problems that limit their full use. Researchers have thus attempted to produce hybrid viruses to exploit the benefits and minimize the disadvantages. For example, an adenovirus–retroviral hybrid packages the retroviral structural genes and the therapeutic gene within the adenovirus (or within two different adenoviruses). The adenovirus can be produced at high titer and efficiently infect cells. The retrovirus is then produced in situ and can infect surrounding cells, resulting in genomic integration of the transgene. Adenovirus is also being evaluated for its use as a carrier for AAV, lentivirus, or even DNA transposable elements. Safety concerns are at the forefront in all experiments involving the recombination of parts from different viruses that are pathogenic to humans.

EMERGING TECHNOLOGIES

Bacteria, in particular *Shigella flexneri,* and viruses, such as baculovirus, feline parvovirus, and measles virus, are being studied for their possible use in gene delivery.

TARGETING GENE DELIVERY

Gene delivery is targeted to enhance specificity, because limiting transgene expression to the target tissue improves safety. Furthermore, by avoiding delivery of the vector to nontarget tissues, efficacy is improved. Targeting can help overcome the many obstacles and impediments to delivery of DNA to the nucleus.

The simplest way to target a tissue is direct administration of the vector to that tissue, and most initial clinical trials took this approach. Viruses or liposomes are directly injected into tumor masses for cancer gene therapy or applied directly to the surface of the respiratory tract for cystic fibrosis. The main advantage of this method is its simplicity. Direct administration avoids some of the barriers to vector delivery, such as the vasculature. Although these approaches may provide a proof of principal, the hope for the future is clearly to develop molecular targeting approaches so that the vector can be given systemically. Receptor-mediated gene delivery involves cell surface receptors that provide an avenue for both targeting gene delivery to the cell surface, entering the cell by endocytosis,and vector internalization into the nucleus.

LIGANDS

To target a specific cell type, the cell in question must either express a unique receptor or overexpress a common receptor. In addition, a ligand for that receptor needs to be identified. Examples of ligand-receptor combinations that have been used for delivering genes are shown in Table 7-1.

TABLE 7-1. LIGAND RECEPTOR COMBINATIONS

Receptor	Ligand	Cell Type	Delivery Vehicle
Asialoglycoprotein receptor	Asialoglycoprotein	Hepatocyte	Polylysine
Transferrin receptor	Transferrin	Several	Polylysine
Epidermal growth factor receptor	Epidermal growth factor	Cancer	
Folate receptor	Folate	Several	Polylysine
Surfactant protein receptor, A and B	Surfactant protein receptor A and B	Airway epithelium	Polylysine
Polymeric immunoglobulin A receptor	Immunoglobulin A	Airway epithelium	Polylysine

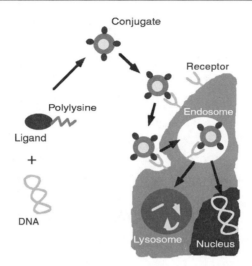

FIGURE 7-9. Ligands are associated with DNA via bridge-like polylysines. When the complexes are delivered to the cell surface, receptor binding and receptor-mediated internalized of the complex can aid gene delivery.

These ligands have usually been associated with therapeutic DNA by using a polylysine bridge. When the complexes are delivered to the cell surface, receptor binding and receptor-mediated internalization of the complex can aid gene delivery (Fig. 7-9). Despite the elegance of this approach, this field of gene therapy has been hampered by poor efficacy of gene expression and DNA degradation within the cell.

ANTIBODIES

Monoclonal antibodies specific for a variety of cellular targets have been explored in several settings. Antibodies directed against a specific target can be added to lipoplex or polyplex complexes to target the vector to the desired cell. Both retroviruses and adenoviruses have been retargeted by constructing bispecific antibodies that bind to both a viral coat protein and to the desired cellular target.

VIRAL PROTEIN MODIFICATION

As noted earlier, viruses have developed many strategies for efficient delivery of their genetic material into the nucleus of the target cell. A primary characteristic is the ability to bind to receptors on the cell surface. Different viruses usurp different cellular receptors. The choice of receptor targeted by the virus influences the repertoire of tissues that can be infected. For instance, the CAR receptor for the adenovirus, which is shared by the coxsackie virus, is expressed on many epithelial surfaces but at low levels on hematopoietic cells (blood forming stem cells). Thus the adenovirus is good vector for targeting the epithelium but a poor vehicle for hematopoietic cells.

The diversity of cells that express receptors for the adenovirus poses a problem for targeting; thus researchers have attempted to change the receptor specificity of the virus to limit infection to certain cells or to enhance infection of cells that express low levels of CAR. The wild-type virus uses the coxsackie adenoviral receptor to mediate binding and an integrin (usually $\alpha v \beta 5$) to mediate internalization into the cell. The virus binds to its receptor via the knob region of the fiber and to the integrin via RGD sequences (arg-gly-asp) within the penton proteins at the base of the fiber. If the knob region of the fiber is modified and the RGD sequences or polylysine sequence is inserted, the virus is no longer limited to infecting only cells that express CAR. This approach decreases the specificity but increases the efficacy of the vector. The fiber knob can also be modified to express a specific ligand, such as the gastrin releasing peptide, so the virus can be targeted to only the cell types that express the receptor for that ligand.

SITE-SPECIFIC REPLICATION

A major limitation in cancer gene therapy is achieving high enough levels of the therapeutic agent within the tumor cell to ensure efficacy (cell toxicity) while maintaining low enough levels in nontarget tissues to ensure safety. The concept of using a virus that could specifically replicate in tumor cells, increasing the local dose several thousandfold, is an exciting proposition.

Several innovative strategies, mainly using adenovirus, have been pursed to achieve this aim of tumor-specific viral replication. One method involves changing the promoter of the adenoviral *E1a* gene, which is essential for viral replication. The native *E1a* promoter is active in all cell types. Replacement of the viral promoter with the promoter of a gene that is preferentially expressed in a tumor—for example, the α-fetoprotein 1 gene promoter for hepatoma or the prostate-specific antigen promoter for prostate cancer—can limit viral replication to the targeted cell type. The target cell can then be killed, either as a result of the lytic viral infections or by transfer of a therapeutic viral gene.

Another approach is to exploit the similarities between viral and tumor cell biology (Fig. 7-10). In essence, both tumor cells

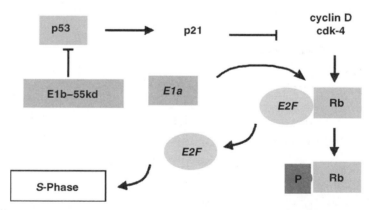

FIGURE 7-10. Two approaches to tumor-specific viral replication. In normal cells, the adenoviral E1b-55kd protein inhibits p53 to enable viral replication. In tumor cells with mutated p53, E1b-55kd is not necessary; thus an E1b-55kd-deleted virus will replicate only in cancer cells. In normal cells, *E1a* releases E2F from Rb, which enables the cell to enter the S phase of cell division, allowing viral replication. In tumor cells with mutated Rb, Rb-binding of *E1a* is not required; thus an *E1a*-modified virus should replicate only in cancer cells.

and adenoviral-infected cells share two common needs: to undergo cell division and to overcome signals for apoptosis (protective cell death) so they can stay alive, despite uncontrolled cell division. This duplication of function can be exploited to target viral replication to tumor cells. The adenovirus needs to block the function of the host tumor-suppressor gene p53, so it can push the host cell into cell division and replicate itself efficiently. However, the majority of tumors already have mutations in the *p53* gene that render the p53 protein nonfunctional. If the virus is modified so that its p53-blocking gene (*E1b*) is removed, it will not be able to replicate in normal cells. However, the *E1b*-deficient virus would be able to replicate in tumor cells that have inactive p53. Clinical trials to support this approach are currently in progress (Khuri et al., 2000). Similar methods focus on mutations in the adenoviral *E1a* gene to target Rb-mutated cells and use of a Reovirus to target *Ras* overexpressing cells.

GENE REPLACEMENT

ADENOSINE DEAMINASE DEFICIENCY

Adenosine deaminase (ADA) deficiency is the cause of severe combined immunodeficiency (the "boy in the bubble" disease). In the absence of ADA, a toxic product accumulates in lymphocytes, resulting in dysfunction of both T and B cells. An approach to restoring a normal copy of the ADA gene to lymphocytes was the first gene therapy trial to negotiate the ethical, regulatory, and safety regulations and to be approved in the United States. The first trial began in 1990 at the National Institutes of Health. Two children received retroviral-mediated transfer of the ADA gene to their lymphocytes, *ex vivo*, over a 2-year period (Grossman et al., 1995), and there is evidence for the persistence of transduced cells over several years. However, these patients also received standard therapy for ADA, so confirming evidence of a clinical response from the gene therapy itself has been difficult. In several subsequent trials, peripheral lymphocytes, bone marrow cells, and umbilical cord blood cells of neonates with ADA deficiency have been transduced. Persistence of the transgene has been frequently seen.

CYSTIC FIBROSIS

Cystic fibrosis (CF) is a homozygous recessive disorder that is the most common inherited disease in the Caucasian population worldwide. The gene that is mutated in CF encodes a chloride channel termed the cystic fibrosis transmembrane conductance regulator (CFTR). The disease is primarily manifest on the respiratory epithelium, which becomes particularly susceptible to bacterial infection. Frequent infections of the bronchi lead to tissue distortion and destruction. Although the exact mechanism by which abnormal gene function leads to disease is unclear, individuals heterozygous for the gene mutation are free from disease, and experimentation has shown that as few as 1 corrected cell in 10 could reverse the abnormal CF phenotype. In addition, the respiratory epithelium (which includes the nasal epithelium) is reasonably accessible, and efficiency of gene transer can be measured repeatedly by molecular techniques as well as by physiologic techniques that ascertain functional correction.

Many protocols have been initiated to attempt a functional correction of the respiratory epithelium in individuals with CF, using delivery methods that include adenovirus, adeno-associated virus, and liposomes. All of the techniques used showed some evidence of gene transfer and functional correction, but all were faced with considerable barriers to achieving effective gene transfer. In addition, one individual developed an inflammatory response within the lung after the administration of adenovirus.

This was the first clinical suggestion that gene delivery, in this case by the adenovirus, may be associated with adverse effects. Furthermore, repeated administration was required, because none of the vector systems resulted in integration of the transgene into target cell chromosomes. This posed particular problems for the first-generation adenovirus, which induced a strong cell-mediated and humoral immune response. Newer generation viruses express much less viral protein and are therfore likely to induce less of an immune response.

FAMILIAL HYPERCHOLESTEROLEMIA

Familial hypercholesterolemia is an autosomal dominant disorder in which the gene encoding the low-density lipoprotein (LDL) receptor is defective. Reduced levels of this receptor result in high levels of circulating cholesterol, which leads to premature atherosclerosis and myocardial infarction. Individuals who are homozygous for the mutation are at risk and have a markedly reduced life expectancy. Because a clinical benefit is noted when this receptor is expressed at low levels ($>10\%$ of normal), research has focused on a genetic correction. Grossman and co-workers (1995), described attempts to partially restore receptor expression in five individuals who were homozygous for the genetic defect. A retrovirus, which would lead to stable integration of the transgene, was targeted to the patients' livers; unfortunately, hepatocytes have a low basal level of replication. To overcome this problem, the researchers obtained liver cells that could be cultured and infected *ex vivo* and then returned to the patients through a portal venous catheter. Gene expression could be detected for 4 months, and the patients experienced no adverse effects. However, the metabolic effects were variable, and subsequent research has been directed to more efficient means of transgene delivery.

OTHER GENETIC DISEASES

Clinical trials have been conducted to test gene therapy approaches for other genetic diseases, including chronic granulomatous disease, ornithine transcarbamylase deficiency, hemophilia B, Canavan disease, mucopolysaccharidosis type 1, Gaucher disease, α-1-antiprotease deficiency, Fanconi anemia, and leukocyte adherence deficiency. Clinical protocols for the following diseases are just beginning: Huntington disease, hemophilia A, gyrate atrophy, muscular dystrophy, Fabry disease, amyotrophic lateral sclerosis, junctional epidermolysis bullosa, JAK3-deficient severe combined immunodeficiency, and purine nucleoside phosphorylase deficiency.

EXPRESSION OF THERAPEUTIC GENES

REVASCULARIZATION

Research has progressed in several directions for developing therapies to maintain or develop a vascular supply. One of the main problems with coronary artery vein grafts, is the subsequent development of neointimal hyperplasia, which narrows the graft lumen and leads to impairment of the vascular supply to the heart. Strategies to prevent the accumulation of new intimal cells include the use of oligonucleotides that block the effects of genes that are important in cell division. A phase I study has begun using an oligonucleotide as an E2F decoy. E2F is an important transcription factor that regulates genes that increase cellular division. The presence of a decoy that mops up the E2F transcription factor might therefore block cellular proliferation.

Attempts have also been made to coat stents that are used for coronary artery stenting with endothelial cells that have been modified to express a therapeutic gene. The therapeutic gene may inhibit clot formation or secrete vascular growth factors to stimulate new vessel formation downstream.

Vascular endothelial growth factor has been administered directly to the myocardium via an adenovirus to stimulate new vessel formation in ischemic hearts. So far, these studies have only confirmed the feasibility of the technique.

NEURONAL DISORDERS

Protocols being explored for Parkinson disease include the transfer of the gene for tyrosine hydroxylase to relieve symptoms and the transfer of antioxidant genes to lessen the neuronal deterioration. For Alzheimer disease, the transfer of nerve growth factor to limit neuronal degeneration has received consideration. A clinical trial using genetically modified xenogeneic cells to express ciliary neurotrophic growth factor has also been reported.

STRATEGIES FOR CANCER THERAPY

Advances in therapy since the 1980s for many solid organ tumors have not lead to substantial improvements in cure rates or survival. Gene therapy provides a potentially new therapeutic modality. The problems faced when developing a therapy for cancer include achieving a sufficiently high dose of the agent within the tumor to enable effective tumor cell death while minimizing toxicity to normal tissues. Many tumors have low levels of immunogenicity, which limits the natural host defense against the tumor. Tumors are genetic diseases that result from multiple gene mutations; some of the mutations may be inherited, but most are acquired. Mutations that increase the function of genes active in tumor cells are called oncogene mutations; such as mutations in the *ras* gene that lead to the unrestrained activation of the RAS proteins. Conversely, tumor-suppressor mutations deactivate genes that represses tumor formation in normal cells (e.g., *p53*). Treatment strategies include modification of the immune response, expression of a therapeutic drug, correction of any acquired genetic mutations within the tumor, and modification of nontumor stem cells to allow the administration of higher does of conventional therapy.

IMMUNE RESPONSE MODIFICATION

Modification of the immune response encompasses two areas (Fig. 7-11). One is to try and make the tumor more immunogenic, and the other is to recruit antigen-presenting cells to the tumor. To induce an immune response, tumor antigens have to be presented to T cells. Although many tumor cells express antigens, they are not efficiently presented. Furthermore, tumor cells usually do not express the appropriate ligands required to activate lymphocytes, and they are not good at recruiting antigen-presenting cells to perform this task. Thus gene therapy vectors have focused on improving the tumor cell's antigenic presentation by transferring an appropriate ligand to the tumor. Gene transfer is accomplished by directly injecting the vector

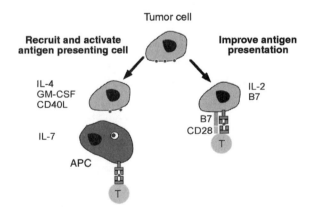

FIGURE 7-11. Strategies to enhance antitumor immunity. Genes can be transferred to tumor cells either to improve antigen presentation or to recruit antigen-presenting cells (APC).

into the tumor or by infecting tumor cells *ex vivo* and then administering the modified tumor cells.

Another approach is to transfer a gene that will activate and recruit antigen-presenting cells to the tumor. Some genes under investigation are those for granulocyte-macrophage colony-stimulating factor, CD40L, and interleukin 1.

GENE THERAPY

Genes encoding proteins that are cytotoxic to cancer cells have been transferred to tumors, including genes for cytokines like tumor necrosis factor and proapoptotic proteins like Bax. Another strategy is to transfer a gene that can activate a prodrug to a toxic product. For example, the thymidase kinase gene, gives the cell the ability to phosphorylate the prodrug gancyclovir to a toxic product. The drug can then be administered systemically, but its toxic effect would act only on the modified tumor cells.

GENETIC MODIFICATION

Gene therapy strategies to inhibit tumor gene (oncogene) expression include using antisense sequences to block the *ras* gene. Attempts to replace the function of the mutated p53 tumor-suppressor gene have used several vector systems and have focused on several different malignancies.

STEM CELLS

Chemotherapy drugs are meant to be specifically toxic to rapidly dividing tumor cells. The side effects of chemotherapy are primarily due to toxic effects on other dividing cells in the body such as stem cells. Bone marrow stem cells are usually quite sensitive to the toxic effects of chemotherapy, and this can limit the administration of effective tumor-killing doses. A gene therapy remedy is to transfer the gene for a multidrug-resistance protein (P-glycoprotein) to marrow stem cells. This treatment is intended to increase the resistance of stem cells to the toxic effects of the chemotherapy to enable the administration of more effective doses of chemotherapy.

THE PROBLEMS

The application of genes as a therapy is faced with the same hurdles as the application of any new medicine. The therapy must have the desired therapeutic effect, and this must be achieved with an acceptable level of toxicity. Most of the gene therapy trials have approached diseases that allow a large therapeutic widow, for which overexpression of the gene is not likely to produce serious adverse effects, and for which a small amount of gene transfer may have beneficial effects. For example, as noted earlier, a correction of 1 in 10 cells in a cystic fibrosis patient is probably enough to provide functional correction to the epithelial layer, and gene overexpression does not appear to be toxic. However, for a disease like diabetes, for which precise regulation of insulin secretion is required, many more hurdles need to be overcome.

Unlike chemotherapy agents, for which the predominate problem is toxicity, the main problem with gene therapy trails is lack of efficacy. The impressive gene transfer and expression that are seen with a variety of vectors for *in vitro* systems have not translated well to human trials. Achieving adequate and persistent levels of gene expression at the target site has proven difficult, and each vector system seems to have its own Achilles' heel. Retroviruses, for instance, are not stable in the circulation, and early-generation adenoviruses induce a robust immune response and are rapidly cleared from the body by immune and nonimmune mechanisms. Transferred genes that have become stably integrated have shown loss of expression over time. This gene silencing probably occurs as a result of promoter methylation.

Toxicity has not been a major problem, with the exception of one or two highly publicized adverse events, the most notable of which was the death of a young man after the administration of an adenoviral vector directly into the portal vein. This was probably due to a systemic inflammatory response induced by the adenoviral vector. This incident lead to a rigorous reevaluation of the conduct of gene therapy trials and the establishment of the Office of Human Research Protection to provide more detailed guidance on the performance and monitoring of compliance of clinical trials.

THE FUTURE

Despite the relatively few therapeutic successes, the first decade of gene transfer technology has provided an enormous amount of scientific and clinical information and built a firm platform for future developments. Initial successes seen with SCID-X1, hemophilia B, Leber congenital amaurosis, and the use of replicating adenoviral vectors for head and neck cancers promise a bright future:

> *SCID-X1.* Mutation of the γ-cytokine receptor subunit of the interleukin-2, -4, -7, -9, and -15 receptors result in a severe combined immunodeficiency (SCID-X1). A gene therapy approach for SCID-X1 that uses retrovirus for ex vivo transfer of the corrected gene to CD34+ cells results in full restoration of function for up to 10 months.
>
> *Hemophilia B.* Intramuscular injection of an AAV vector expressing factor IX into dogs with severe hemophilia B results in a dose-dependent increase and prolonged expression of

circulating levels of the factor, This level of expression is enough to provide phenotypic improvement in humans. A clinical trial using humans has been initiated.

Leber congenital amaurosis. Leber congenital amaurosis causes near total blindness in infancy and can result from mutations in the *RPE65* gene. A naturally occurring animal model, the *RPE65-/-* dog, suffers from early and severe visual impairment similar to that seen in affected humans. A recombinant AAV vector carrying wild-type *RPE65* restored visual function in dogs.

Replicating adenoviral vector. A replicating *E1b*-deficient adenoviral vector targeted to tumor cells with mutated p53, used in combination with chemotherapy, induced some complete response of injected lesions in head and neck cancer.

The main research focus in the early twenty-first century is likely to be on achieving adequate levels of gene expression in the absence of an inflammatory response to achieve meaningful modulation of human disease. The positive aspects of different vector systems are likely to be combined and used in combination with more traditional therapies. The enormous international scientific interest in gene therapy is clearly evident by the rapid expansion of professional groups like the American Society of Gene Therapy and of meetings and journals devoted to the field.

REFERENCES

Avery OT, Macleod CM, McCarty M. Studies on the chemical nature of the substance inducing transformation of pneumococcal types. J Exp Med 1944;79:137–158.

Blaese RM, Culver KW, Miller AD, et al. T lymphocyte-directed gene therapy for ADA-SCID: Initial trial results after 4 years. Science 1995;270:475–480.

Griffith F. The significance of pneumococcal types. J Hygiene 1928;27:113–159.

Grossman M, Rader DJ, Muller DW, et al. A pilot study of ex vivo gene therapy for homozygous familial hypercholesterolemia. Nature Med 1995;1:1148–1154.

Khuri FR, Nemunaitis J, Ganly I, et al. A controlled trial of intratumoral ONYX-015, a selectively-replicating adenovirus, in combination with cisplatin and 5-fluorouracil in patients with recurrent head and neck cancer. Nature Med 2000;6:879–885.

SUGGESTED READING

Acland GM, Aguirre GD, Ray J, et al. Gene therapy restores vision in a canine model of childhood blindness. Nature Genet 2001;28:92–95.

Cavazzana-Calvo M, Hacein-Bey S, de Saint Basile G, et al. Gene therapy of human severe combined immunodeficiency (SCID)-X1 disease. Science 2000;288:669–672.

Crystal RG, McElvaney NG, Rosenfeld MA, et al. Administration of an adenovirus containing the human CFTR cDNA to the respiratory tract of individuals with cystic fibrosis. Nature Genet 1994;8:42–51.

High KA. Gene transfer as an approach to treating hemophilia. Circ Res 2001;88:137–144.

Mann MJ, Whittemore AD, Donaldson MC, et al. Ex-vivo gene therapy of human vascular bypass grafts with E2F decoy: The PREVENT single-centre, randomised, controlled trial. Lancet 1999;354:1493–1498.

Riordan JR, Rommens JM, Kerem B, et al. Identification of the cystic fibrosis gene: Cloning and characterization of complementary DNA. Science 1989;245:1066–1073.

MICROARRAYS

The growth of the term *genomics* both among biologists and in the popular media has largely been driven by the development of microarray technology for the measurement of gene expression. The concept of a gene array is quite simple: A large number of DNA sequences for known genes (targets) are attached in defined locations to a surface, creating an array of spots. Many different surfaces have been used for DNA arrays, but glass microscope slides coated with something that enhances the covalent binding of DNA have become the standard (Fig. 8-1). (The glass slides bearing these spots of DNA are often called "DNA chips"; the term *GeneChip* is a trademark of Affymetrix Corp.) A labeled test sample of total RNA sequences extracted from some cells (the probe) is then applied to the surface. Probe sequences bind by RNA–DNA hybridization to complementary target sequences in the array. Then the amount of labeled RNA bound to each target spot is measured. The amount of RNA probe bound to each target reflects the amount of that RNA sequence in the test sample and, therefore, the level of expression of that gene. A microarray is simply a gene array in which the DNA targets are applied in very tiny spots.

Essentials of Medical Genomics, Edited by Stuart M. Brown.
ISBN 0-471-21003-X. Copyright © 2003 by Wiley-Liss, Inc.

FIGURE 8-1. A DNA microarray on a glass slide. (Reprinted with permission from Ferea and Brown, 1999.)

MEASURING GENE EXPRESSION

Microarrays represent an extension of two older molecular biology techniques: the measurement of gene expression by using hybridization to quantify the amounts of specific mRNAs (e.g., Northern blots), and the measurement of many genes at once in a dot blot hybridization. Microarrays are a genomic technology, because the concept of a dot blot has been scaled up to allow measurement thousands of genes in parallel. Ideally, a microarray would contain a target for every gene in a genome. This has been essentially achieved for the ~6000 genes in yeast (DeRisi et al., 1997), but the genome sequences of humans and other higher eukaryotes are not yet sufficiently complete to make a DNA probe for every gene.

Microarray experiments are always conducted as a comparison between two (or more) samples. The relative amount of probe bound to each spot in the array is compared between the two samples, and a ratio is calculated. This ratio may be expressed in absolute terms or, more typically, in terms of a fold change—so that a given gene might be observed to increase in expression 2.5-fold when the target tissue is subjected to an experimental condition as compared to the control condition. A different gene will undergo a different relative change in expression level across the same two samples. The comparison of two samples may be done by making separate measurements of labeled RNA on two identical arrays or by labeling two samples with different colors (different fluorescent makers) and mixing them together in a single hybridization reaction on a single array (Shalon et al., 1996). Even if two samples are mixed and hybridized together, the two different probe colors are measured independently by a fluorescence scanner at different wavelengths; then the two images are combined to create a false-color image. Typically, red is used to represent one probe and green the other. Spots that show up as red or green in the combined image indicate a significantly higher level of mRNA for that gene in one sample than in the other. Yellow spots indicate high levels of expression in both samples and dark spots indicate low expression in both samples (Fig. 8-2).

This measurement of RNAs extracted from a tissue sample is actually a snapshot measure of mRNA transcript abundance. It is not exactly the same as a direct measure of gene expression— the actual amounts of each gene product (i.e., protein) being manufactured in the cell at a given moment in time, nor is it a

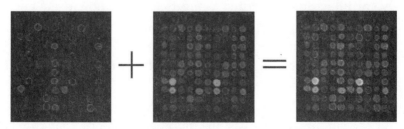

FIGURE 8-2. Two separate fluorescent microarray (with red and green false colors) are combined to show the relative gene expression in the two samples. Figure also appears in Color Figure Section.

direct measurement of DNA transcription into mRNA. Other factors (e.g., different translation efficiencies, different rates of degradation of mRNA species, and alternative splicing) play roles in modulating gene expression. However, microarray measurements of mRNA are usually well correlated with both levels of transcription and levels of protein abundance in the cell.

Gene expression microarrays composed of DNA targets (DNA chips) are intended to report the transcriptional levels of all genes to investigate metabolic differences caused by disease, development, and other variables. However, alternative splicing of mRNA must be taken into account in the design of these chips. At a minimum, alternate splice forms can confuse results because they may show differential binding to probes that bind to **exons,** which are sometimes spliced out. If the knowledge of alternate splice forms is incorporated into the DNA chip design, then expression of alternate forms can be directly measured and incorporated into the overall **gene expression profile.**

SPOTTING VS. SYNTHESIS

The array hybridization technique is useful for making rough quantitative measurements of the current expression levels of a bunch of genes at once; but as originally conceived, it had several limitations. The first being that only known sequences could be used in the array. That limitation is much less of a problem now that the human genome has been completely sequenced as well as the genomes of many bacteria, yeast, *Caenorhabditis elegans, Drosophila* spp., and mouse; and the complete genomes of many other species are soon to come. The second limitation was the number of genes that can be practically handled—both in the process of building the array and in measuring the binding of labeled sequences. That

FIGURE 8-3. A microarray spotting robot making many chips at once.

limitation has been vigorously and rather successfully addressed by several different technical innovations. Genomics is all about the automation of biology and conducting experiments in a massively parallel fashion. Robotic fabrication technologies developed for other industrial applications are well suited to the precise and repetitive work of spotting tiny amounts of DNA onto glass slides (Schena et al., 1995) (Fig. 8-3). Automated data collection and analysis software has also been developed to allow high-throughput gene expression measurements with microarrays.

These robotically assembled arrays can contain almost unlimited numbers of tiny spots of DNA, so they have become known as microarrays. Currently, there are two fundamentally different types of microarray technologies and many variations on these types. There are a number of robotic systems designed to place spots of cloned **cDNA** or polymerase chain reaction (**PCR**) amplified fragments onto a substrate—either a glass slide or a nylon membrane. **Plasmids** carrying clones of cDNA are generally maintained as frozen stocks, and purified plasmid DNA is stored in microtiter plates. Then each clone is amplified by PCR to make DNA fragments for spotting onto chips by the spotting robots. This allows researchers to determine the expression level of thousands of different genes simultaneously and with high sensitivity. RNA from two different samples—a control and an experimental—are each labeled with different colored fluorescent dyes. These labeled RNAs are combined and hybridized together onto a single array. The fluorescence of each label is measured separately, and the ratio of the two is calculated for each spot (Fig. 8-4).

Alternately, it is possible to synthesize short DNA oligomers of specific sequence directly onto known locations on a grid. Affymetrix Inc. developed a system called GeneChip™ that uses photolithographic technology (similar to that used in the manufacture of computer chips) to synthesize hundreds of thousands of different DNA oligomers simultaneously on a single chip (Lockhart et al., 1996) (Fig. 8-5). The light-activated DNA synthesis is guided by a series of masks with holes that either allow

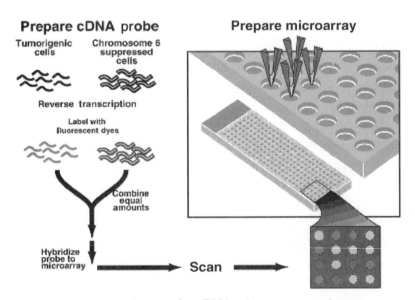

FIGURE 8-4. A two-color cDNA microarray experiment.

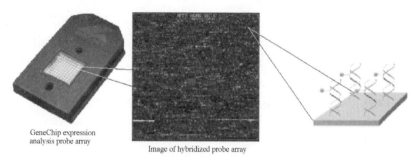

FIGURE 8-5. The Affymetrix GeneChip with one cell enlarged to show the labeled probe bound to the oligonucleotide targets. (Reprinted with permission from the Affymetrix website www.affymetrix.com.)

or prevent a given base to be added to the oligonucleotides in each location on the chip.

Each gene is represented by 16 to 20 different 25-base-long oligonucleotide probes that cover the length of the coding region. In addition, for each probe, a second "mismatch" probe is added that has a single base mutation in its center (Fig. 8-6). Affymetrix has developed software that calculates the ratio

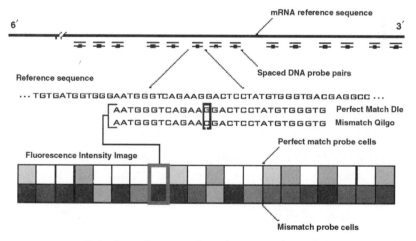

FIGURE 8-6. Paired perfect match and mismatch probes used in an Affymetrix GeneChip. (Reprinted with permission, *Nature Genetics*, from Lipshutz et al., 1999.)

between the binding of labeled RNA to the perfect match probe and the mismatch probe and combines the ratios across all 20 probes to generate a single value to represent the level of mRNA for each gene and an estimate of the quality of that measurement. To make a comparison between two treatments, the two RNA samples must be hybridized to separate but identical GeneChips™.

An intermediate between using short 25-base targets on the array (Affymetrix method) and using full-length cDNA as targets exists. *Long oligos* are 40- to 80-base DNA segments from each gene of interest that are synthesized as oligonucleotides and then spotted onto an array. The selection of target sequences depends heavily on bioinformatics methods. First, a single representative DNA sequence must be generated for each gene—a synthesis of all available information from cDNA and genomic sequences. Then, a single section of the sequence must be chosen that is specific to the expressed part of that gene (i.e., lacks significant sequence similarity with any other gene) and that adheres to a variety of primer design criteria (e.g., optimal GC content, optimal melting temperature, and lack of self-complementary sequences).

There is clearly a trade-off between sensitivity and specificity for targets of varying length. Full-length cDNA targets bind more labeled RNA of different lengths and all splice variants, so they can be more sensitive for detecting signals in smaller RNA samples and less common RNA species. However, cDNA probes also bind RNA from homologous genes, such as the members of a multigene family, so they are less specific. The short, 25-base oligo probes in Affymetrix GeneChips™ bind RNA less efficiently, so they are less sensitive, but they can be more specific for one member of a gene family and even for one particular splice variant. Long oligos, if designed well, can combine good sensitivity with the ability to discriminate among similar genes and splice variants.

There are also pragmatic trade-offs in the choice of cDNA, short oligos, and long oligo microarray targets. The Affymetrix method of creating a microarray chip requires a labor–intensive process of designing custom probes for each gene and then creating a series of photolithographic masks to allow simultaneous synthesis of all of the oligos on a chip. It is quite cumbersome and costly to change just a few of the oligos on a chip or to design a custom chip with a new set of genes, so individual scientists rarely create their own GeneChips™ for experiments that require less than thousands of chips. Chips created from spotted cDNAs are much more flexible in terms of the ability to change a few genes or create a custom set of genes on a chip for a few experiments. However, when thousands of cDNA probes are used on a chip, the management of the clones, production of DNA for spotting by PCR reactions, verification, and quality control become challenging for an academic laboratory. Affymetrix GeneChips™ offer the most efficient way to survey the expression of entire genomes. Long oligos offer the flexibility to change a few targets in an array or to create a custom array; but since there are no clones to grow and no PCR reactions to prepare, there are fewer potential sources of experimental error in chip construction than for cDNA chips. Some vendors of molecular biology reagents are now offering to produce sets of custom oligonucleotides in microtiter format ready for use by chip spotting robots or to sell chips spotted with the investigator's choice of long oligos, which may be custom designed from private sequences, chosen from a list of public genes or from predesigned groups.

OTHER TYPES OF ARRAYS

There are many other applications of microarray technology besides the measurement of gene expression. Microarrays can be built that determine the sequence of a specific fragment of DNA according to its differential binding to targets that contain variant sequences at each position. These arrays can be designed to provide the entire sequence of a specific fragment of DNA or simply to detect sequence variants at a single known position on a fragment (as for a single nucleotide polymorphism; **SNP**). Arrays for the detection of SNPs can also be reversed so that fragments of DNA from specific genes of interest are amplified from patient samples by PCR and are bound as targets on a chip. Then probes of variant sequences, each carrying a different fluorescent label, are hybridized to the array. The sequence of the target gene in each patient (the SNP genotype) is determined by the color of the probe that binds to each spot on the array.

It is also possible to build arrays of proteins as targets (protein chips)—either by directly attaching purified proteins to a substrate or by attaching colonies of cells that produce the desired proteins from cloned expression vectors. Then it is possible to use these proteins as targets for labeled probes that might consist of other proteins (to look for protein–protein interactions) or of fragments of genomic DNA (to assay DNA-binding properties). Antibodies could be attached to a chip as targets in an array. Then specific proteins could be quantitated from a labeled mixture according to the amount bound to each type of antibody. The problem with protein arrays, however, is that unlike mRNA, there is no single universal biochemical approach that can provide global, genome-wide profiles for all different types of proteins—they are just too chemically diverse (see Chapter 10).

DIFFERENTIAL GENE EXPRESSION

The microarray is a flexible tool for measuring gene expression in any tissue sample that can be isolated for RNA extraction. The primary research application of this technology has been the study of differential gene expression as the result of development, disease, or the response to chemical agents such as drug treatments. A set of such measurements of gene expression levels across many (or all) genes for a set of treatments is a *gene expression profile* that provides important functional information about many genes. From a broad collection of such profiles created for many different tissue types under many different experimental conditions, an understanding of *gene regulatory networks* will emerge.

It is important to keep in mind that microarray technology measures the *relative* changes in the expression of many individual genes from one treatment to another. It does not measure absolute quantities of RNA from each gene, nor does it allow for comparisons of absolute amounts of gene expression from one gene to another in a single sample. There are many unknown factors that may influence the relative efficiencies of a highly diverse set of molecules in solution (the labeled mRNA from the samples) when binding to a diverse population of DNA molecules attached to the spots in the array.

Simple microarray experiments aim to identify differences in gene expression (amounts of mRNA from different genes) across a single treatment or tissue type: cancerous vs. normal, drug treated vs. untreated, etc. Data analysis for this type of experiment usually amounts to looking for genes with significant changes in expression levels between the two samples (red vs. green or more than a twofold change) and making sure that this change is reproducible when the experiment is repeated (Fig. 8-7).

More complex experiments might characterize a time course or a set of related tissues, such as subtypes of a particular form of cancer. These more complex data sets contain patterns of expression that are unique for each gene. The challenge is to group the genes into clusters based on similar patterns of changes of the gene expression values across the various samples (Fig. 8-8). Many different clustering techniques have been

FIGURE 8-7. A spotted cDNA array hybridized with a mixture of two probes and two different fluorescent labels visualized as a red–green false-color image. Figure also appears in Color Figure Section.

applied to microarray data sets, but no single analysis method has yet emerged as the best. There is also a challenge in weighing the importance of relative changes in RNA amount (up threefold from sample A vs. sample C) and absolute changes (an increase in fluorescence from 1000 to 1500 vs. from 10 to 200) as well as the need to incorporate statistical tests to ensure the validity of these changes.

Microarray measurements are especially useful for identifying correlations between the expression of various genes—gene expression clusters. These patterns of co-regulation correspond to metabolic pathways and coordinated cellular responses to developmental and environmental cues. It is expected that all of the genes in a metabolic pathway, such as synthesis of the amino acid tyrosine, would be turned on and off together. In contrast,

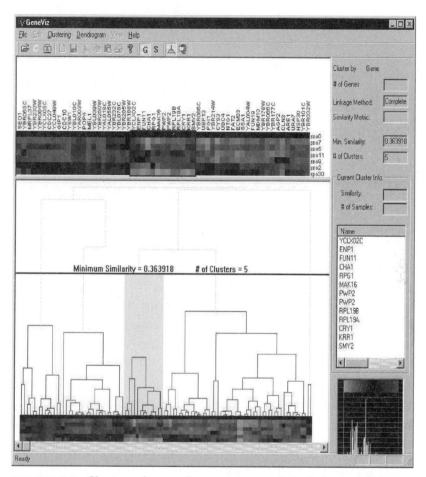

FIGURE 8-8. Clusters of genes that are expressed similarly over different experimental treatments. (Reprinted with permission from Seo and Lee, 2001.) Figure also appears in Color Figure Section.

the members of a protein family, such as kinases, are involved in many different metabolic pathways that are subject to different regulatory controls. A common characteristic of microarray experiments is that clusters of co-regulated genes with previously described functions are found that confirm some hypotheses about dynamic cellular processes, but other clusters are observed that can not be explained by current biological theories;

and thus serve to stimulate fresh thought and new insights into fundamental biological processes.

In each set of co-regulated genes identified in a microarray experiment, there are likely to be some genes that have been previously studied and assigned a function as well as some previously unknown genes. In addition, some genes with known functions will be regulated in unexpected ways. By collecting information about differential regulation under a variety of conditions and co-regulation with other genes, a detailed picture can be developed for the expression and regulation of every gene.

At the present time, microarray technology is running far ahead of our knowledge of gene functions. It is possible to create arrays based on cDNA sequences or predicted genes from genomic sequence, but after the experiments are completed and clusters of genes with interesting changes in expression levels are found, nothing useful may be discovered about these genes in public databases. In fact, this is an iterative process. The results of microarray experiments are themselves useful bits of annotation that should be captured in genome databases— tissue-specific, disease-related, and drug-responsive expression are important aspects of gene function. Then as new genes are found to be co-regulated in common processes, new functional motifs can be defined. Then new drugs may be found to interact with the proteins encoded by these genes (or that modify gene expression), closing the circle from clinical phenotype to molecular genetic analysis and back to clinical therapy.

CLASSIFICATION BY GENE EXPRESSION

It is also possible to use the coordinate expression of a particular set of genes as a marker for a cellular process or a disease state. For example, various types of cancer are usually diagnosed by a microscopic examination of a tissue biopsy by an expert pathologist (histology), but these diagnoses are not always 100% accurate. This is extremely important for the patient, because histologically similar types of cancer with different cellular and genetic origins often respond differently to treatment and have dramatically different prognoses. It is possible to assay samples of these different types of cancer tissues and identify genes that consistently have levels that differ between the cancer types. A set of these diagnostic genes on a microarray chip can then be used to categorize an unknown tissue sample with extremely good reliability. This is known as a **class prediction.** The most meaningful set of genes for characterizing a sample into a group may be much smaller than the set of genes that show differential expression. Ideally, genes used for class prediction should consistently differ in expression between the two classes and should have a low variance within a class. It is interesting to note that genes that serve to reliably classify samples are often useful as drug targets or provide important insights in the cellular mechanisms of disease.

Golub and co-workers (1999) developed a microarray class prediction method to distinguish between acute myeloid leukemia (AML) and acute lymphoblastic leukemia (ALL). They used 50 genes (out of a total of 6817) on an Affymetrix GeneChip^{TM} to reliably classify a set of unknown samples (Fig. 8-9). One sample diagnosed by classical methods as atypical AML did not match the class prediction for either form of leukemia, but an examination of some highly induced genes suggested a muscle origin, which was confirmed as rhabdomyosarcoma by cytogenetic analysis.

There are a number of diseases, such as prostate cancer, for which there is no good diagnostic standard. A microarray-based assay that could clearly distinguish prostate cancer from other types of variations in prostate size and could differentiate elevated levels of prostate-specific antigen (PSA) caused by ordinary inflammation or hyperplasia instead of by cancer

B   **ALL**   **AML**

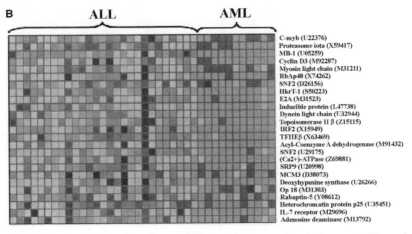

FIGURE 8-9. Genes that show differential expression between ALL and AML. (Reprinted with permission Science Magazine, from Golub et al., 1999.)

would save lives and prevent a large number of unnecessary surgeries and aggressive radiation and chemotherapies. A further characterization of tumors in terms of their aggressiveness and responsiveness to various drugs would revolutionize prostate cancer therapy. Furthermore, a good diagnostic would be an important asset in clinical drug trials—allowing researchers to limit trials to patients with aggressive tumors. A change in gene expression profile might also serve as an early indicator of treatment success.

Microarray experiments sometimes reveal additional subcategories of cancers that cannot be distinguished histologically. These subcategories may correlate with different responses to drugs or other treatments, so the more precise the diagnosis made from a microarray the greater the direct benefit to the patient. Alizadeh and co-workers (2000) used microarrays to study diffuse large B-cell lymphoma. By clustering gene expression results, they found that tissue samples fell into two classes that represented different stages of B-cell differentiation. The two groups also correlated with patient survival rates.

Once sets of genes have been well defined and verified to predict types of cancer cells or for other diagnostic purposes, the technology could be broadly commercialized. DNA chips are an expensive research tool right now, but they could be mass-produced cheaply. It is important to understand the difference between a gene expression–based diagnostic test and a classic genetic test that examines a patient's genomic DNA. Genetic tests measure genomic DNA, which is unchanging and identical in every cell of the body. Gene expression tests measure the current activity in a small sample of cells—that is: what proteins are currently being made in a specific tissue. It can be used to differentiate cell types or to measure the health of a group of cells. Gene expression tests are sensitive to differences in the environment, the patient's metabolic status, and the manner in which the sample is collected and processed.

Microarrays are particularly useful in basic research and the early stages of drug development. By studying what genes are induced and repressed in diseased vs. normal tissue, key functions can be inferred and potential drug targets, identified. The up- or down-regulation of a gene may indicate that it is a cause or a result of disease processes, but in either case, a drug that returns that gene to normal levels may provide a beneficial therapy.

ERROR AND RELIABILITY

Before microarray technology can be used for routine medical diagnostics, a variety of issues related to sampling accuracy must be resolved. Real patient tissues, such as tumor biopsies, contain a mix of normal and diseased cell types, including nerve cells, blood cells, immune cells, vascular tissues, and connective tissues. Most patients have multiple tumors with different levels of aggressiveness and with different gene expression profiles. Even cells or tissues that appear histologically normal may be in early stages of developing cancer, or their gene expression may be influenced by nearby cancerous cells. Microarray methods must be robust enough to reliably detect key diagnostic markers in mixed or impure samples.

Microarray experiments produce a large quantity of numerical data, which are the output of fluorescent scanning of the labeled RNA bound to the spots in the array. However, the readout of the fluorescent sensor is an indirect measurement of the amount of each gene's mRNA in the corresponding sample. There are many possible sources of measurement error, including RNA extraction and labeling, hybridization, irregularities in the scanning process, and image processing (finding the boundaries of each spot and integrating its total fluorescent signal, subtracting a background value). Without dwelling excessively on the technicalities of these issues—which are shared by many other data acquisition technologies—it is important to realize that microarray data require some type of standard error calculation. The technique relies heavily on the ratio of gene expression between experimental and control treatments, but such a ratio can be misleading for spots with low fluorescent intensities (i.e., for genes that are expressed at low levels in the sampled cells).

The only way to estimate the reliability of microarray measurements is through the use of replicates. However, GeneChips^{TM} are expensive; and in many studies with pathology samples or microdissected tissues, it is extremely difficult to obtain enough material for a statistically valid number of replicates. In addition, it requires sophisticated software to analyze microarrays in order to accommodate reliability and standard error measurements for hundreds of thousands of data points.

Affymetrix GeneChip^{TM} experiments do contain some internal validation in the ratios of signal in perfect-match to mismatch probes, but little has been done to use this information for improved measurements of reliability for each gene's expression level. Similarly, the fluorescent image of each spot on a cDNA array contains additional information about the shape and uniformity of the signal within that spot, but again little has been done to generate a reliability value for each gene from this information.

Another issue in the interpretation of microarray data is the relationship between the probe on the array and the actual mRNA in the sample. Our current knowledge of human genes is incomplete. Many more cDNA (and expressed sequence tags; **EST**s) exist in the databases than the number of genes predicted in the genome, so if a set of microarray probes are made from these database cDNAs, there is not a complete correspondence with well-defined genes. Whether the probe is a cloned cDNA or oligonucleotides, there are many possible mRNAs that could hybridize. There are sets of closely related genes that share long regions of nearly perfect sequence homology that can produce mRNAs that will cross-hybridize with the probes. These related genes may undergo quite different pattterns of regulation under tissue-specific, developmental, disease, drug-response, and/or other condition. Also, individual genes may undergo many different forms of alternative splicing, which leads to different mRNAs. These alternate transcripts may fail to bind to a probe in an array or may bind with different properties—particularly to sets of oligonucleotide probes. As genome information accumulates, microarrays can be designed to compensate for this complexity of mRNA populations, perhaps even to quantitate amounts of alternately spliced transcripts. However, for now, these must be seen as a source of error and confusion.

Different samples and replicate samples measured on different days will give different values in a microarray experiment. Some of this variation is simply due to systematic changes in experimental conditions—a slightly more efficient buffer or a longer hybridization time, different room temperature, etc. These systematic variations can affect all of the values measured on a chip in the same way—everything is brighter or darker. This can be corrected by normalization of the data (i.e, scaling the data from each chip so that the midpoint is set to a common value). It might seem logical to use the values of some well-established "housekeeping" genes that are known to maintain a steady expression level across all experimental conditions; but after much controversy, no such universally constant genes have been proven to exist. Alternately, a DNA sequence that does not correspond to any mRNA in the sample may be included in the array, and the sample can be spiked with a known amount of a matching sequence before it is labeled. This creates a positive control that can be used to normalize the intensities of the other spots measured on the array. Unfortunately, small variations in the effective concentration of spiked probes can create substantial artefactual differences in the expression levels measured for other genes.

EVOLUTIONARY PERSPECTIVES

Changes in gene expression may be a more subtle and flexible form of evolutionary change than the mutations in protein-coding regions that we have become familiar with in medical genetics. Humans and mice have approximately 92% identical protein sequences, humans and chimpanzees are >99% identical. Data from microarray and other gene expression studies suggest that the larger and more essential differences are in the regulation of *expression* of some important genes not in their sequences. An experiment by Pääbo (at the Max Planck Institute, Germany) revealed that the differences in relative expression levels of 20,000 genes in humans vs. chimpanzees were much more pronounced in the brain than in the blood or liver (Enard et al., 2002). Intuitively, it seems clear that one can build a larger brain or a smaller brain from the same basic set of molecular components by following a slightly different developmental program, which can be modified by changes in gene expression. Its not what genes you have that matters, but how you use them!

> Much of the information encoded in the genome is devoted not to specifying the structure of the protein or RNA that the gene encodes, but rather to controlling precisely in which cells, under what conditions, and in what amounts the gene product is made. Differences in the program of gene expression as opposed to variation in the encoded products may underlie much of the phenotypic variation within and between species.
>
> *Ferea and Brown (1999:715)*

Differences in gene expression may be a common source of evolutionary adaptation in response to selective pressures such as pathogens. Microarrays are better suited to detecting and characterizing such variation than tools such as Northern blot or quantitative PCR, which are limited to one or a few genes at a time. In many cases, changes in expression of genes that are not part of an investigator's initial hypothesis may turn out to be important in a given biological system.

REFERENCES

Alizadeh AA, Eisen MB, Davis RE, et al. Distinct types of diffuse large B-cell lymphoma identified by gene expression profiling. Nature 2000; 403;503–511.

DeRisi JL, Iyer VR, Brown PO. Exploring the metabolic and genetic control of gene expression on a genomic scale. Science 1997;278:680–686.

Enard W, Khaitovich P, Klose J, et al. Intra- and interspecific variation in primate gene expression patterns. Science 2002;296:340–343.

Ferea TL, Brown PO. Observing the living genome. Curr Opin Genet Dev 1999;9:715–722.

Golub TR, Slonim DK, Tamayo P, et al. Molecular classification of cancer: Class discovery and class prediction by gene expression monitoring. Science 1999;286:531–537.

Lipshutz RJ, Fodor SPA, Gingeras TR, Lockhart DJ. High density synthetic oligonucleotide arrays. Nature Genet 1999;21(1 Suppl):20–24.

Lockhart DJ, Dong H, Byrne MC, et al. Expression monitoring by hybridization to high-density oligonucleotide arrays. Nature Biotechnol 1996;14: 1675–1680.

Schena M, Shalon D, Davis RW, Brown PO. Quantitative monitoring of gene expression patterns with a complementary DNA microarray. Science 1995;270:467–470.

Seo J, Lee B. dynamic visualization of gene expression. www.cs.umd.edu/ class/spring2001/cmsc838b/Project/Lee_Seo. Accessed 2/1/2002.

Shalon D, Smith SJ, Brown PO. A DNA microarray system for analyzing complex DNA samples using two-color fluorescent probe hybridization. Genome Res 1996;6:639–645.

PHARMACOGENOMICS AND TOXICOGENOMICS

PHARMACOGENOMICS

One of the first spin-offs from the Human Genome Project (HGP) to reach the practicing physician will be genetic tests designed to aid in prescribing drugs. This technology, known as **pharmacogenomics,** promises to be both simple and relatively noncontroversial. Pharmaco*genetics* is the study of how genes affect the way people respond to medicines. All patients want to receive the most effective drugs that will have the fewest side effects, but up to now there was essentially no information for helping the physician decide which drug would be best for a specific person. It has been known for some time that genetic factors influence the efficacy and side effects of a particular drug in an individual patient, but the physician generally had no way to measure these factors in advance and to take them into account when writing a prescription. Genomics technology promises to make this information easily accessible.

Pharmacogenomics is generally defined as the use of DNA sequence information to measure and predict the reaction of

Essentials of Medical Genomics, Edited by Stuart M. Brown.
ISBN 0-471-21003-X.   Copyright © 2003 by Wiley-Liss, Inc.

individuals to drugs. The theoretical basis for this technology is quite straightforward. There are many proteins that are known to enhance or block the action of specific drugs—through either direct chemical action on the drug molecule (degradation or activation), interaction with a common target molecule (e.g., to block drug binding to a receptor), or regulation of a metabolic pathway that affects drug function. There are also genes that have been shown to cause drug side effects (e.g., non-target receptors that bind to a drug). It is now possible to use single nucleotide polymorphism (**SNP**) markers to identify the alleles of these drug-interaction genes in test populations and then screen patients for the markers before prescribing the drug.

EXAMPLES OF GENETIC TRAITS FOR DRUG RESPONSE

There are many examples of drug–genetic interactions that have been discovered through the unfortunate experiences of people receiving certain drugs. It has been estimated that >2 million people are hospitalized each year in the United States due to adverse reactions to drugs that were properly prescribed, but without knowledge of each patient's unique genetic makeup.

In World War II, the U.S. Army discovered that 10% of African Americans have polymorphic alleles of glucose-6-phosphate dehydrogenase (G6PD) that leads to hemolytic anemia when they are given the antimalarial drug primaquine. Approximately 0.04% of all people are homozygous for alleles of pseudocholinesterase that are unable to inactivate the anaesthetic succinylcholine, leading to respiratory paralysis.

About 10% of the Caucasian population is homozygous for alleles of the cytochrome P450 gene *CYP2D6* that do not metabolize the hypertension drug debrisoquine, which can lead to dangerous vascular hypotension (Kuehl, et al, 2001). There are many polymorphic alleles of the *N*-acetyltransferase (NAT2) gene that cause reduced (or accelerated) ability to inactivate the drug isoniazid. Some individuals developed peripheral neuropathy in reaction to this drug. Some alleles of the NAT2 gene are also associated with susceptibility to various forms of cancer. This is an important point to which I will return in Chapter 11 in the discussion of ethical implications of genomic testing—a test for one trait may reveal information about other genetic factors, through either pleiotropic effects of a single gene or alleles of linked genes.

In other cases, drugs are less effective for people with a specific genetic trait. Patients homozygous for an allele with a deletion in **intron** 16 of the gene for angiotensin-converting enzyme (ACE) showed no benefit from the hypertension drug enalapril, whereas other patients did benefit.

THE USE OF SNP MARKERS

In all of these examples, the drug-response phenotype is associated with a specific allele of a single gene. Once that gene and its sequence variations are identified, it is possible to construct a single-gene test, or in some cases a biochemical assay, for the variant protein. These are approaches that were available to twentieth-century geneticists. In the era of medical genomics, the identification of pharmacogenomic traits can proceed much more rapidly, and multigene effects can be identified almost as easily as single-gene traits. It is possible to use a panel of thousands of SNP markers that cover the entire genome to screen groups of patients receiving a specific drug and then to correlate good and poor drug efficacy and the occurrence of specific side effects with individual SNP markers (or groups of markers). These linked markers can be used directly to predict drug-response traits or they can be used as landmarks on the genome to initiate a second round of more precise screening with additional sets of SNP markers that are focused on particular chromosomal regions. Then, without ever identifying the genes involved in the process, the linked SNP markers can be used to predict the efficacy and likely side effects of the drug on new patients.

In some cases, differences in drug response between groups of patients identified by pharmacogenomic screening may indicate fundamentally different disease mechanisms. In other words, people with similar symptoms may be experiencing different disease subtypes and thus require different treatment. This is especially likely in complex diseases such as asthma or heart disease, for which a pharmacogenomic test may reveal similar data to a genetic test based on inherited risk factors. In fact, by using pharmacogenomics to divide patients into subtypes, it may be possible to develop new drugs that specifically benefit only one subclass of patients.

DRUG DEVELOPMENT RESEARCH

Drug companies are also using pharmacogenomic technology to speed up the clinical trails process for new drugs (the most expensive and time-consuming phase of dug development). It is possible to make genetic profiles of patients in the early stage clinical trials of a drug and correlate these profiles with drug response and side effects. Then for later stage trials, patients can be prescreened to eliminate those likely to respond poorly or to experience side effects. The result of such stratified trials will probably be drugs that are approved for use only in conjunction with the genetic test that determines if it will be effective and safe for each patient. While this represents some loss of profit compared to a "one drug fits all" marketing strategy, the drug companies will more than make up for it by the ability to license many drugs that were previously disqualified due to low levels of efficacy on some people or unacceptably high frequencies of side effects in other people. When used together with genetic testing, drugs will be safe and useful for some people. In addition, new targeted drugs will be sold to niche populations who previously did not benefit from drug treatment.

At a more prosaic level, there are often a number of different drugs available to treat a given condition—high blood pressure, anxiety/depression, migraine, etc. In the current health-care system, a patient might receive a prescription for one of these drugs based on whatever their physician has read lately about the incidence of side effects, known negative drug interactions, etc. After taking the drug for some period of time, the physician assesses the effectiveness of the drug and the severity of the side effects and then decides whether to continue the prescription or to change to another drug that may be more suitable for the patient. In this way, the patient may suffer for many weeks or months (or longer) with one or more ineffective drugs and/or unpleasant side effects when a better drug was available all along, if only the physician had more information about that patient's genetic drug-response characteristics.

Pharmacogenomics can provide this type of information and can help the physician in determine appropriate drug dosages. Current methods of basing dosages on weight and age will be replaced with dosages based on a person's genetics—how effective the medicine is in that person's body and the time it takes to metabolize it. This will maximize the therapy's value and decrease the likelihood of overdose.

GENETIC PROFILES VS. GENE EXPRESSION

Pharmacogenomics is based on matching drugs to patients according to genetic profiles—identifying specific alleles of known genes or SNP markers linked to these alleles in each patient. These are permanent characteristics of the genome of each person. There are other situations in which it is not the genotype of the patient that determines the effectiveness of a drug but rather the metabolic state of a particular affected tissue. Different types of cancer tumors respond differently to chemotherapy agents, but it is often difficult to diagnose a tumor accurately with classic histologic pathology methods. However, the gene expression patterns of different tumor types can be distinguished using microarrays that measure levels of mRNA for various genes (see Chapter 8). This has been demonstrated convincingly for acute myeloid leukemia (AML) vs. acute lymphoblastic leukemia (ALL) (Golub et al, 1999). In other cases, a gene expression profile of a tumor can accurately indicate its aggressiveness, which can be useful in determining appropriate drugs and a course of treatment, such as for prostate cancer.

PERSONALIZED MEDICINE

Pharmacogenomics is often described as personalized medicine or designer drugs, but these terms encourage a misunderstanding of the basic technology. Pharmacogenomics will not involve designing a drug specifically for each patient. Instead, pharmacogenomics offers a form of "mass-customization" of drugs, so that the physician can choose from among a panel of available drugs the one best suited to each patient—sort of like a choice of size and colors in a sweater, not like a custom-made suit of clothes.

The National Institute of General Medical Sciences (NIGMS, a branch of the National Institutes of Health), is currently funding a major research initiative called the Pharmacogenetics Research Network. Research conducted by scientists in the network includes identification of important genetic polymorphisms, functional studies of variant proteins, and studies that relate clinical drug responses to genetic variation. NIGMS is creating a free online database of pharmacogenomic information collected by the scientists and physicians participating in the program (Pharmacogenetics Knowledge Base; www.pharmgkb. org/ Klein et al., 2001). This online database is hosted and managed by the Stanford Medical Informatics (SMI) group, in the Department of Medicine of the Stanford University School of Medicine. As of January 2002, the online database contained information about 430 genes that had been shown to affect drug response in clinical studies.

This database is intended to be used as a research tool to help scientists understand how genetic variation among individuals leads to differences in reactions to drugs and to contribute to the development of new genetically targeted drugs. However, the database is freely accessible on the Web by any physician, patient, or interested member of the public. However, the database is organized for research uses, and it does not support simple **queries** by drug name—and even if a gene–drug interaction is listed in the database, there is not likely to be a commercially available genetic test for that gene.

The Pharmacogenetics Knowledge Base includes health information, such as history of disease, physical and physiologic characteristics (height, weight, heart rate, and blood pressure) as well as pharmacogenomic information about any drugs being taken, data on physiologic responses to drugs, and DNA sequences suspected to play a role in these drug responses. However, all personal identifying information has been stripped from these data.

In conjunction with the Pharmacogenetics Knowledge Base, the NIGMS has also created a public education Web site called Medicines for You (www.nigms.nih.gov/funding/medforyou. html). This is an excellent resource for patients who are curious about the potential for personalized medicine.

ENVIRONMENTAL CHEMICALS

Just as people have genetic differences in their responses to drugs, they also have genetic differences in their responses to toxic chemicals that occur as environmental pollutants (or as food contaminants, food additives, etc.). Here again, it is possible to collect genetic data on people who demonstrate specific sensitivities to chemicals. The National Institute of Environmental Health Sciences (NIEHS) has initiated the Environmental Genome Project to systematically identify common sequence polymorphisms in genes with suspected roles in determining chemical sensitivity.

The NIEHS has set up an online database of genetic data linked to susceptibility to environmental chemical exposure. The NIEHS database—GeneSNPs, developed and hosted by the University of Utah Genome Center (www.genome.utah.edu/genesnps/)—contains human genes and sequence polymorphisms related to DNA repair, cell cycle control, cell signaling, cell division, homeostasis, and the metabolism of environmental chemicals. This database is freely available to anyone on the Web, but it is oriented toward the researcher with a specific gene in mind rather than to the physician or layperson with an interest in a particular chemical and its possible genetic effects. However, the display of data for each gene is quite impressive—perhaps the best integrated resource for any collection of genetic data. Each gene is shown with its name, functional category, coding regions, introns, 5' and 3' untranslated regions (UTR, part of the mRNA that is not translated into protein), and an additional 10 kilobases of genomic DNA that flanks the coding region on both sides. All known SNPs are annotated as to their location on the gene (5'UTR, 3'UTR, intron, or **exon**). For SNPs that occur in an exon, the database further notes if the mutation is synonymous (e.g., CAG 229 CAA :: Gln 229 Gln) or nonsynonomous (e.g., GTA 163 ATA :: Val 163 Ile) and lists the amino acid position in the translated coding sequence (Fig. 9-1).

These data on genetic sensitivity to toxins could be used to help people make lifestyle choices to avoid certain chemicals for which they have a genetic sensitivity. However, it could also be used to discriminate against people, for example, when making

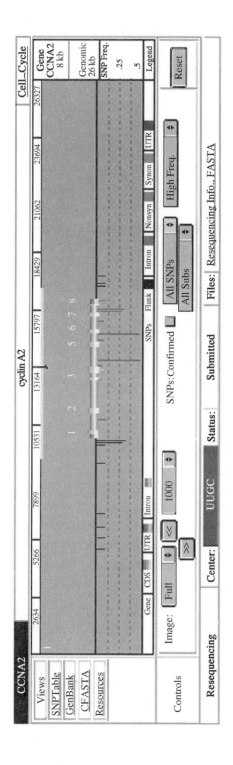

View Frame

FIGURE 9-1. Genetic data for the cyclin A2 gene in the GeneSNPs database. Reprinted with permission from the GeneSNPs Public Internet Resource, NIEHS; www.genome.utah.edu/genesnps.

193

hiring decisions to limit employer liability for on-the-job exposure to potentially toxic chemicals. This is an ethical gray area—
we don't want people to be needlessly exposed to chemicals for
which they are particularly sensitive, but we also do not wish to
live in a genetic dictatorship where the results of a gene test are
used to limit or determine our employment options (Rothstein
and Epps, 2001).

Another ethical concern is the overlap of these various
genetic tests. It is entirely possible, even likely, that if a person
were to have a simple genetic test taken to predict the best
medicine for a common disorder—perhaps for a prescription for
high blood pressure medicine—that test would also reveal
information about chemical sensitivity, cancer risk factors, etc.
Control over ownership, privacy, and the right to permanently
and completely destroy this type of genetic information are going
to be crucial factors in determining the widespread adoption of
this technology in routine medical practice (see Chapter 11).

TOXICOGENOMICS FOR DRUG DEVELOPMENT

DNA microarray technology can be used to measure differences
in gene expression (mRNA levels) due to any type of developmental change, response to a pathogen, or environmental sti-
mulus. This is particularly useful in drug-development research
where it might be used to detect toxic responses to molecules
that are being evaluated as potential drugs. The majority of drug
toxicity reactions (adverse side effects) occur in a few well-
defined biochemical pathways—common modes of action.
These toxic reaction pathways involve induction and/or
down-regulation of specific sets of genes. Microarray technologies can be used to detect these changes in gene expression,
provided that RNA can be extracted from samples of affected
tissues. The gene expression changes observed on a microarray

due to the reaction to a specific toxic drug form a **toxicogenomic** profile both for that drug and for that pathway of toxic response.

The early stages of testing drug candidate molecules usually involves various types of cell culture systems, followed by animal studies (usually mice). Many pharmaceutical researchers are currently working to establish cell culture systems that accurately reflect human toxic reactions to drugs. In such a system, gene expression profiles can be created for various compounds known to have toxic side effects. Then each new drug candidate can be applied to the cell culture, and the resulting gene expression profile can be compared with the library of known toxicogenomic profiles. Similarly, tissue samples can be collected from affected tissues in mice suffering from known toxic reactions and microarray gene expression profiles compiled. Samples of the same tissues can be collected from mice being treated with experimental drugs and tested on the microarrays. If the gene expression profile of a new drug candidate matches a known toxic profile, then that candidate molecule is eliminated from further testing, saving time, money, and possibly the suffering of unlucky drug-testing animals or humans.

It is also possible to use microarray screening to monitor patients who are currently receiving drug treatment—either as an early alert for toxic effects or to assay drug effectiveness. The crucial distinction between toxicogenomic technologies and standard genetic tests is that it is the *present state* of cellular metabolism that is being measured not the permanent genetic characteristic of the patient. A drawback of this technology is that samples must generally be collected directly from the tissues that are the subject of the drug treatment (e.g., a liver biopsy).

DRUG SPECIFICITY

A drug with effects narrowly targeted to a single tissue is much less likely to produce unwanted side effects than a drug that produces effects on many different tissue types. Tissue-specific drug effects could be monitored by microarray tests of mRNA extracted from various tissues in an experimental animal treated with the drug. However, this expression profile is only as specific and precise as the tissue used to prepare the RNA, since tissue samples may be composed of many different cell types, which can be difficult to separate.

Toxic effects of drugs are sometimes linked to their mode of action—unanticipated metabolic side effects of blocking or stimulating a particular protein or receptor too well—but more often, they are the result of interactions between the drug and nontarget molecules. A new genomic technology has been proposed to use microarrays to specifically screen for nontarget effects of drugs. Marton and co-workers (1998) (at Rosetta Inpharmatics, Kirkland, WA) have proposed a method for investigating nontarget effects of drugs. For several drugs, they have created gene expression profiles in a wild-type strain of yeast and then compared these to gene expression profiles of a yeast strain with a mutation in the primary target gene affected by the drug (or another gene in the same metabolic pathway). The concept is that the mutant strain will show little effect when exposed to the drug, since its target is absent—thus validating that gene as the true drug target. Furthermore, any nontarget effects of the drug will be much easier to detect in the mutant strain when the usual metabolic effects of the drug are absent. Thistechnology is quite powerful in yeast, which has only 6000 well-characterized genes, and producing specific mutants is a simple laboratory procedure. The procedure is much more challenging in mice, and essentially impossible (with current technology) in humans.

ENVIRONMENTAL TOXICOLOGY

There are also environmental and toxicological applications for microarray technology. Just as gene expression profiles can be established for drugs with toxic effects, similar profiles can be established for various toxic chemicals that might occur in the environment or be used in industry. These profiles could then be used to establish the mechanism of action for new chemicals with suspected toxic effects.. This is important, considering that there are approximately 80,000 chemicals in commercial use in the United States, and an additional 1,000 new chemicals are developed each year.

There is no way that industry can afford to conduct complete batteries of animal tests for all of these chemicals, nor could the Environmental Protection Agency scrutinize complete animal testing results. A preliminary toxicogenomic study of a chemical could be accomplished quickly and inexpensively—perhaps on a small set of well-defined cell cultures. A streamlined preliminary testing procedure based on microarrays would fit in nicely with the current law known as the Toxic Substances Control Act (TSCA), which requires that a premanufacture notification (PMN) be submitted for each new chemical. Currently, the PMN is not required to contain any toxicity data, which has come under criticism from environmental advocacy groups. Adding toxicogenomic data to the PMN would be both inexpensive and informative.

Toxicogenomic profiles might also be used diagnostically to help determine what type of chemical might be causing an adverse health effect in a person for whom exposure to toxic chemicals is suspected. Gene expression microarrays could provide a more sensitive and earlier assay for exposure to toxic chemicals than current blood chemistry or physiologic tests. Similarly, in a situation where a specific toxic chemical has been released (or suspected to have been released) into the environment, gene expression testing could provide a sensitive measurement of each person's degree of exposure.

REFERENCES

Golub TR, Slonim DK, Tamayo P, et al. Molecular classification of cancer: Class discovery and class prediction by gene expression monitoring. Science 1999;286:531–537.

Klein TE, Chang JT Cho, MK, et al. Integrating genotype and phenotype information: An Overview of the PharmGKB Project. Pharmacogenomics J 2001;1:167–170.

Kuehl P, Zhang J, Lin Y, et al. Sequence diversity in CYP3A promoters and characterization of the genetic basis of polymorphic CYP3A5 expression. Nature Genet 2001;27(4):383–391.

Marton MJ, DeRisi JL, Bennett HA, et al. Drug target validation and identification of secondary drug target effects using DNA microarrays. Nature Med 1998;4:1293–1301.

Rothstein MA, Epps PG. Ethical and legal implications of pharmacogenomics. Nature Rev Genet 2001;2:228–281.

PROTEOMICS

Proteomics is a hot new buzzword surfacing in scientific conferences and journal articles as well as among biotechnology investors. Proteomics can be loosely defined as the measurement and study of all of the proteins in an organism (or in a specific tissue or cell type)—that is, the **proteome.** This covers the full gamut of information about proteins from amino acid sequences to tissue-specific expression, three-dimensional structure, protein–protein and protein–DNA interactions, and biochemical and metabolic function. This is closely associated with **functional genomics,** which seeks to understand the function of all of the genes in the genome.

Gene expression, as measured by microarray technology provides some information about where and when genes are expressed, but most genes exercise their biological function through the production of proteins. Proteins are the enzymes, regulatory molecules, and building blocks of cellular structures. Unlike DNA microarrays, which apply a uniform technology based on RNA–DNA hybridization to measure the mRNA produced by all genes, proteomics involves an assortment of different technologies. Proteomics technologies include

Essentials of Medical Genomics, Edited by Stuart M. Brown.
ISBN 0-471-21003-X. Copyright © 2003 by Wiley-Liss, Inc.

quantitative measurements of different molecular species of
proteins (mass spectroscopy), identification of protein–protein
interactions, protein structural analysis, two-dimensional gel
electrophoresis, and various forms of computational function
prediction. The essential distinction between classic protein
chemistry and molecular biology approaches and the new
proteomics methods is that proteomics attempts to address all
of the proteins in an organism at once.

PROTEIN MODIFICATIONS

Proteomics is significantly more complex than DNA- or mRNA-
based genomics technologies. While alternate splicing allows
one gene to produce several different mRNAs, **post-translational modification** of proteins can multiply this complexity
many fold. Proteins can be cut by specific proteases (proteolytic
cleavage), cross-linked by disulfide bonds either internally or to
other protein molecules (of the same type or of different types),
phosphorylated, glycosylated, hydroxylated, carboxylated, ami-
dated at the C-terminal, acylated, and methylated. Sulfate can be
added to tyrosine residues, and farnesyl or geranyl groups
added to carboxy terminal cysteine residues. Complex carbo-
hydrate molecules are linked to some proteins. Proteins may
also be localized within the cell, excreted from the cell, transported through the body, and bound at cell surface receptors or
imported by specific cell types that are located far from the cells
that produced them. Proteins can also bind to other proteins or
to nonprotein cofactors to form complex molecular machines,
such as ribosomes, membrane pores, and spliceosomes. All of
this complexity leads to the inevitable conclusion that identifying and measuring all of the proteins from cells or tissues is
going to produce an extremely large data set of chemically
diverse molecules.

These post-translational modifications have a profound in-fluence on the function of a protein in a specific cellular system. Proteins with the same amino acid sequence but in different cross-linking, phosphorylation, or glycosylation states may have different metabolic activities. Ideally, investigators wish to know the exact amounts of every form of every protein as a precise measurement of the biological state of a tissue or cell type.

QUANTITATIVE APPROACHES

Quantitative measurement of proteins potentially provides the most precise information about a patient's current heath and metabolic status. Genomics experiments, which use DNA micro-arrays to measure RNA levels, can make only approximate measurements of the levels of proteins being produced from each gene. RNAs are the template for protein synthesis, but there are many forms of **post-transcriptional regulation** that can affect the amount of protein made from each mRNA molecule. The mRNAs for some genes may be translated into protein more efficiently than others. This may be due to direct sequence-specific differences in the processing of mRNA sequences by the ribosomal translation machinery or due to the actions of capping and other mRNA-processing enzymes, which may themselves have sequence-specific affinities. Alternately, some mRNA molecules may be degraded by RNAse enzymes more rapidly than others and, therefore, may serve as a template for the synthesis of fewer protein molecules. There may also be a significant time lag between changes in mRNA levels and changes in the overall levels of the corresponding proteins— large pools of some proteins may exist in the cell, buffering the effect of a rise or fall in the rate of new synthesis. In fact, some proteins may have long lifetimes in a cell, so that rates of gene transcription and protein synthesis may not be closely related to the amounts of that protein in the cell.

Measuring the amounts of many different proteins in a cellular extract is technically much more difficult than measuring mRNA in DNA microarrays because proteins are so chemically diverse. Proteins range from strongly acidic to strongly basic, hydrophilic to hydrophobic, membrane bound or soluble and may be glycosylated, attached to metallic or organic cofactors, bound into dimers, or made part of complex multiunit molecular machines. Any measurement technology that isolates proteins from a cellular lysate must favor some chemical forms over others. Every buffer and reagent will have differential effects on this complex mixture of molecules. There is no single technology that can capture and quantitate all (or even most) of the proteins produced by a cell. Therefore, proteomics technologies will inevitably be a composite of many different biochemical methods.

Current methods for measuring proteins involve some combination of gel electrophoresis, chromatography, affinity binding, and mass spectrometry. Two-dimensional polyacrylamide gel electrophoresis (2-D PAGE) has been the traditional workhorse of protein chemistry for several decades. A sample of proteins (cellular extract) is first separated by pH in one direction in an acrylamide gel by isoelectric focusing. Then proteins are separated by size in a second dimension in the same gel (at right angles to the first) by electrophoresis. Finally, proteins are visualized in the gel by staining or autoradiography. If all experimental conditions are kept constant, then the same protein should end up in the same location on the gel in repeated experiments, so that samples from different tissues or experimental manipulations can be compared to see if the amount of protein present in a specific spot increases or decreases. This involves quite complex image-analysis software since *many* proteins are present and no two gels are ever exactly alike (Fig. 10-1). Even with the most perfectly controlled gels and excellent image-analysis tools, the intensity of protein spots on a 2-D PAGE can provide only a rough estimate of protein amounts.

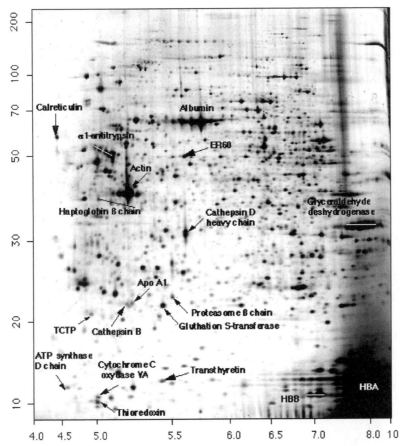

FIGURE 10-1. A 2-D PAGE image of proteins from human lymphoma tissue.

Once proteins are separated by 2-D PAGE, it is possible to identify them with a variety of techniques. Individual proteins can be identified by immunoblotting: transferring the proteins from the acrylamide gel onto a nylon membrane and then using a specific antibody to bind the protein of interest and some immunostaining technique to identify where on the gel the antibody binds. Individual spots can also be cut out from the gel so the proteins can be identified by mass spectrometry or amino acid sequencing. A number of databases have been created to facilitate the analysis of 2-D PAGE that identify the pH and mass locations of various proteins and offer many images for comparison; but the identity of every protein spot must still be confirmed.

Proteomics requires a high-throughput method to identify and quantitate huge numbers of different proteins in parallel, rather than one at a time. New mass spectrometry technologies such as matrix-assisted laser desorption/ionization time of flight (MALDI/TOF) mass spectrometry offer the most promise. MALDI/TOF works by ionizing protein molecules with a laser, accelerating the ions in an electric field, and then measuring the amount of time required to reach a detector. The mass of a protein molecule is proportional to its time of flight to the detector, and its amount is proportional to the intensity of signal recorded at the detector at a particular moment in time. Proteins can be uniquely identified by the mass signatures of their ionization products. MALDI/TOF can identify a number of mixed proteins, even in a tiny and impure sample (femtomole quantities), and the machine has rapid throughput. However, it still cannot sort out tens of thousands of proteins in a heterogeneous mixture. Some separation technology must be applied first.

A variety of affinity-based separation schemes are currently in use to separate proteins from a cell extract before MALDI/ TOF analysis. Any one of a number of generic chemical or biochemical affinity ligands can be used to bind entire classes proteins by their intrinsic biochemical properties. Proteins can also be labeled in solution with a linker molecule; then that linker is used as an affinity tag to attach the proteins to a substrate. It is also possible to use specific antibodies to fish out individual species of protein molecules from a mixture or to create tagged proteins by gene fusions, which can later be retrieved by affinity binding to the tag.

To make a high-throughput, genome scale experiment, it would be necessary to screen all proteins as targets. This has been partially accomplished with the protein chip by placing a huge number of different proteins in an array on a glass slide. First, the gene for each protein is cloned into an expression vector; next, the resulting protein product is purified and placed in a spot on the array. Then the cell extracts are washed over the array, and the protein-protein interactions are screened in an all-against-all format. So far, these protein arrays have been used mostly to measure the interactions of each of the proteins on the chip with a single substrate, such as calmodulin, streptavidin, or phosphatidylinositide. The protein chips can also be used to characterize the interactions of an array of proteins with a drug molecule. The combination of protein chips with MALDI/TOF should soon lead to a technology that can identify and quantitate every protein (or a least a significant fraction of the proteins) in a cellular extract.

USES FOR QUANTITATIVE PROTEIN DATA

Quantitative protein data can be used for many of the same applications as mRNA-based gene expression data: identification of co-regulated proteins; determination of tissue-specific (or subcellular) localization of proteins; identification of quantitative differences in proteins associated with specific genotypes or phenotypes; and association of changes in the abundance of specific proteins in response to disease, drugs, or toxic substances (i.e., proteomic signatures that can be used as diagnostic tools). All of these measurements could potentially be more precise than mRNA-based technologies because they reflect the actual amounts of proteins active in the cells and because it is possible to discriminate between various forms of a protein (phosphorylation state, post-transcriptional modifications, etc.).

PROTEIN DATABASES

It is a primary goal of the human genome project to create a single, definitive list of all of the genes and all of the expressed proteins in the human genome and to assign functions to each of these proteins. However, in 2002, this goal seems quite distant. There are dozens of protein databases that rely on various interpretations of genomic data, each containing some entries that are not shared by the others. The National Center for Biotechnology Information (NCBI) has created a "hand curated" list of human proteins known as RefSeq NP, which currently has 14,039 entries (January 2002). The NCBI also maintains a list of predicted proteins from the public genome sequencing project (RefSeq XP), which has 31,327 entries. **Ensembl** (a genome annotation effort maintained by the European Molecular Biology Laboratory) contains 28,706 human proteins, but many of them are different from the proteins listed in RefSeq XP. **SwissProt** is a high-quality manually annotated database of protein sequences maintained by the Swiss Institute of Bioinformatics (SIB), which contains 7,652 human proteins (out of a total of 103,370 proteins from all species). The SIB also maintains a more comprehensive nonredundant list of proteins called Translations of the EMBL DNA database (**TrEMBL**), which has 32,299 entries, again substantially different from the RefSeq and Ensembl lists. The European Bioinformatics Institute has developed a cross-referenced protein database for all of these others known as the International Protein Index (IPI). As of January 2002, it contained 51,925 entries.

PROTEIN-PROTEIN INTERACTIONS

Another aspect of proteomics is the physical interaction between proteins. Many proteins interact with other proteins—to form complex multisubunit molecular machines to regulate the function of other proteins, or to be regulated. The extent and the nature of these interactions are important for understanding metabolic and regulatory pathways and for the functional characterization of the many new proteins being discovered by genome sequencing (functional proteomics). Many proteins form complex multisubunit structures that may include two or more copies of the same protein (homodimers or homo-polymers) or complexes with other proteins (heterodimers or multimeric structures). These multisubunit structures can assume the complexity of full-fledged molecular machines, such as ribosomes, histones, DNA and RNA polymerases, and the RNA splicing complex.

Several different technologies are available for examining protein–protein interactions, but none has the capacity for high-throughput analysis of the complete protein complement of a cell or an organism. The traditional biochemical method of investigating protein–protein interactions is to attach a purified protein (the target protein) to a matrix, such as a resin in a chromatographic column, and then to pour a cellular extract over the matrix, allowing some proteins to adhere by binding to the target protein. Then the bound proteins would be chemically characterized—generally by mass spectroscopy.

The yeast two-hybrid system improved this method by allowing the protein–protein interaction to take place inside a yeast cell; clones of genes for each protein that bound to the target protein were selected. Even so, the mapping of all inter-actions between all proteins would require a separate experi-ment using each different protein as the target.

A recent innovation in protein–protein interactions has been the use of gene fusions to add affinity tags to the ends of cloned genes for a large number of individual proteins (Gavin et al., 2002; Ho et al., 2002). These tagged proteins can then be expressed in yeast cell lines, where they are used as "bait" for other interacting proteins. Under a given set of metabolic conditions, the tagged protein plus any other proteins bound to it are collected on an affinity column; then the captured proteins are separated from the tagged bait protein and identified by mass spectroscopy. So far these technologies have been applied only in yeast, in which the construction of thousands of transgenic strains with affinity tagged protein sequences is quite simple. In principle, however, they could be applied to any cells that can be grown in culture.

Protein–protein interactions and metabolic pathways can also be predicted computationally. One approach to this problem is to use the evolutionary tendency for multiple proteins that function in sequential steps in a metabolic pathway to become fused into a single gene in some organisms. For example, the α and β subunits of the fungal tryptophan synthetase gene correspond to two separate bacteria genes. Similarly, a set

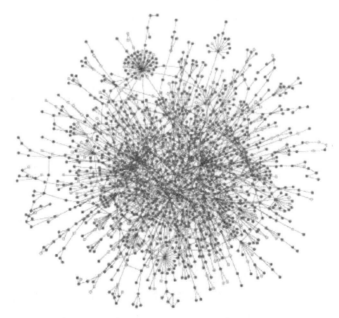

FIGURE 10-2. A map of protein–protein interactions for 1870 yeast proteins. (Reprinted with permission, *Nature* from Jeong et al., 2001, Macmillan Publishers Ltd.) Figure also appears in Color Figure Section.

of proteins that function in a single metabolic pathway tend to be conserved across evolution—it is unlikely that a species would keep some members of the pathway and discard others. Thus, as more complete genome sequences accumulate for more species, the functional annotation of human proteins can be filled in. Computational predictions of protein interactions can be combined with experimental data and with knowledge of biochemical and signal transduction pathways to form protein-interaction maps (Fig. 10-2; color insert). These maps can then be used as a tool to validate or interpret genomic and proteomic results that indicate co-regulation.

DNA BINDING PROTEINS

Many proteins regulate the transcription of other proteins by binding to the genomic DNA near the coding sequence of the regulated protein and interacting with the RNA polymerase machinery. These **transcription factors** bind to specific DNA sequence motifs that are generally located in the promoter region—directly upstream from the transcriptional start site. Other transcription regulators known as **enhancers** bind elsewhere on the chromosome, as much as 20 or 40 kilobases from the coding sequence. A given DNA-binding protein may stimulate or repress transcription, or it may have both functions, depending on its protein–protein interactions with other transcription factors, which may themselves bind to other DNA motifs. The control of gene expression seems to be the result of a complex combinatorial interaction of sets of DNA-binding regulatory proteins with the promoter sequences of a gene.

A variety of molecular techniques, such as gel-shift electrophoresis and affinity chromatography, have been used to isolate proteins that bind to a given piece of DNA. Then the individual proteins can be identified using mass spectroscopy. It is also possible to identify the particular sequence to which each transcription factor binds using techniques such as DNAse protection, mutagenesis, and protein-DNA cross-linking. In fact, these transcription factor–binding sites are generally only about 6 bases long, which is not nearly long enough to allow transcription factors to target only the promoters of a precise set of genes. (A 6-base sequence will occur by chance once in every 4096 bases of DNA.) However, promoter sequences contain multiple transcription factor–binding sites, sometimes all for the same factor, but often for two or more different factors. Therefore, the binding of each transcription factor protein to DNA may require cooperative interactions with other transcription factors. The overall interaction of all the transcription factors with all the binding sites in the promoter region of a gene allows the promoter to serve as a complex switching mechanism that is sensitive to small shifts in levels of any of the regulatory proteins. Other possible modes of regulation might include alternate splice forms or chemically modified (phosphorylated, glycosylated, etc.) forms of the transcription factor proteins that might bind DNA more tightly or serve as nonfunctional competitive inhibitors.

A set of mutually interacting transcription factors may function to regulate many different genes that show similar gene expression patterns, such as the members of a single biochemical pathway. However, these same DNA-binding proteins, with the addition of a few alternates, may also regulate proteins in totally different pathways with different tissue-specific, developmental, and temporal gene expression characteristics. Thus a relatively small set of transcription factors serve in various combinations to regulate precisely the expression of many genes.

STRUCTURAL PROTEOMICS

It is expected that a great deal more can be learned about the biological function of a protein from its three-dimensional (3-D) structure than from its amino acid sequence or molecular weight. However, it is not currently possible to predict 3-D structures directly from amino acid sequence or mass spectrometry data. 3-D structures can be determined only experimentally from the painstaking work of purifying and crystallizing a protein and then subjecting the crystal to X-ray crystallography or Nuclear Magnetic Resonance (NMR) analysis. However, it is possible to use information about known protein structures to predict the structures of similar proteins. This process, known as **threading,** starts by looking for protein sequence similarity between a new protein and all of the protein with known structures in a database called the Protein Data Base (PDB), which is maintained by the Research Collaboratory for Structural Bioinformatics (Westbrook et al., 2001). If a similarity is found in the PDB, then threading software can try to fold up the new sequence into a shape similar to that given in the database, making allowances for some known folding properties of specific amino acids that differ between the two proteins.

Threading is limited to proteins that have amino acid sequences that are about 30% identical. Also, when proteins are compared for sequence similarity, only a portion of the two proteins may be similar (Pieper et al., 2002). These conserved regions often correspond to functional domains or motifs, which are actually distinct substructures of the complete protein. In the language of the PDB, they are protein folds. It is interesting to note that these functional folded substructures generally correspond to exons in the genomic DNA sequence, validating Gilbert's (1978) hypothesis that introns allow functional portions of genes to recombine.

The PDB currently contains about 17,000 structures, but these break down into only about 2,500 unique folds. For comparison, the InterPro database recognizes about 4,700 protein domains, but in fact the discrepancy is greater. Each protein family in InterPro contains proteins that have more than 30%

sequence diffferences with each other, so more than one experimentally determined structure is needed for each recognized
functional domain. While efforts are being made to increase the
rate at which protein structures are analyzed, there are significant obstacles to automating and scaling up this process. Crystallizing proteins is still much more of an art form than an
industrial procedure; each protein requires its own optimal pH,
salt concentrations, and the presence of various biological and
inorganic co-factors. Furthermore, the proteome is complex—
each gene may produce multiple protein isoforms due to alternate splicing and/or post-translational modifications, and each
isoform is likely to have a unique structure.

DRUG TARGETS

The application of genomics technologies to developing new
drugs is all about targets. A drug target generally refers to a
protein in the human body that can be acted on by a drug to treat
disease. Until recently, pharmaceutical companies created drugs
against a total of about 500 drug targets. With the help of genomics, this number could eventually balloon to 4000 drug targets.
Microarrays are particularly well suited to identifying many
new genes that are induced (or down-regulated) during the
various stages of a disease. The next step is validating these
potential targets: Is the protein involved in a biochemical or
regulatory pathway that is directly involved with the cause of
the disease of interest (or in the manifestation of symptoms)?
Then comes the search for a drug molecule that can interact with
this protein target to make a useful change—such as to block
disease development or to ameliorate symptoms. If the 3-D
structure of a target protein is known (or can be computationally
predicted), then databases of small molecules can be tested
initially by computer-simulated docking to screen for potential
drugs. Again, this can speed the process of searching for a drug.

However, the most time-consuming steps in the drug development process are the late-stage testing on animals and humans. First, it must be proven that the new drug is safe and has a useful therapeutic effect and then it must be shown that the new drug is more effective than existing treatments. Unfortunately, these steps will not be substantially shortened by proteomic technologies.

REFERENCES

Gavin AC, Bosche M, Krause R, et al. Functional organization of the yeast proteome by systematic analysis of protein complexes. Nature 2002;415;141–147.

Gilbert W. Why genes in pieces? Nature 1978;271:501.

Ho Y, Gruhler A, Heilbut A, et al. Systematic identification of protein complexes in Saccharomyces cerevisiae by mass spectrometry. Nature 2002;415:180–183.

Jeong H, Mason S, Barabási AL, Oltvai ZN. Centrality and lethality of protein networks. Nature 2001;411:41–42.

Pieper U, Eswar N, Stuart AC et al. MODBASE, a database of annotated comparative protein structure models. Nucleic Acids Res 2002;30:255–259.

Westbrook J, Feng Z, Jain S, et al. The Protein Data Bank: unifying the archive. Nucleic Acids Res 2002;30:245–248.

THE ETHICS OF MEDICAL GENOMICS

Part of the rationale for U.S. government funding for the Human Genome Project (HGP) through the National Institutes of Health (NIH) and the Department of Energy (DOE) since 1990 includes setting aside a full 5% of the funds for investigations into the ethical, legal, and social implications of the project (ELSI, 2000). This strong concern over the social impact of genomic research is well founded on historical precedent. For genomic medicine to be accepted by the public, modern genetics must overcome a checkered history that includes many examples of misapplication of genetics to social policy, particularly in the pseudoscience of eugenics. Since current public attitudes toward medical genetics and genomics are shaped by this history, it is necessary to examine it in some detail before addressing the current debate about the ethics of medical genomics and its affect on the medical professional.

Essentials of Medical Genomics, Edited by Stuart M. Brown.
ISBN 0-471-21003-X.   Copyright © 2003 by Wiley-Liss, Inc.

EUGENICS

In the late nineteenth and early twentieth centuries, Sir Francis Galton and Charles Davenport developed the concepts of eugenics as a social policy to improve the human race by encouraging the most genetically fit people to have more children and to prevent reproduction of people deemed unfit. Eugenics was enthusiastically adopted by many respected scientists and by the U.S. government, leading to the establishment of the Eugenics Record Office at Cold Spring Harbor Laboratory from 1910 to 1939.

Eugenics policies were implemented in the United States, ranging from restrictions on immigration to the involuntary sterilization of jailed criminals or persons institutionalized for reasons of "insanity or feeblemindedness." Harry Laughlin of the Eugenics Record Office published a "model eugenical sterilization law" in 1914, which became the basis for laws in 33 states. For example, the state of Virginia enacted the Eugenical Sterilization Act in 1924 based on Laughlin's model law, which stated that "heredity plays an important part in the transmission of insanity, idiocy, imbecility, epilepsy and crime." The law provided for the sterilization of individuals who were "probable potential parents of socially inadequate offspring." Involuntary sterilizations were upheld by the U.S. Supreme Court in 1927 (*Buck v Bell*). Chief Justice Oliver Wendell Holmes wrote in his opinion: "Society can prevent those who are manifestly unfit from continuing their kind.... Three generations of imbeciles are enough" (quoted in Lombardo, 1985).

Eugenic sterilizations were inflicted on >60,000 people in the United States before the last of the laws were repealed in the 1970s (Cold Spring Harbor Laboratory [CSHL] eugenics archive). Eugenics concepts were also incorporated in the "racial hygiene" social policies of the German Third Reich government. In fact, the German government directly adopted text from

Laughlin's model law as the basis for its 1933 Law for the Prevention of Defective Progeny, which was used as the legal basis for the sterilization of >350,000 people. The Nazi policy of racial cleansing led to the mass extermination of millions of Jews, Gypsies, homosexuals, and other "disfavored" groups.

Eugenics was broadly accepted through American society during the 1920s and 1930s. There were eugenics exhibits at state fairs, contests for "fitter families," films, public lectures, religious sermons, and eugenics chapters in biology textbooks. In the scientific establishment, there were university courses, eugenics foundations, societies, newsletters, conferences, and scholarly journals (Fig. 11-1).

This history of eugenics seems misguided from our current perspective, since our present understanding of genetics contradicts the vast majority of the eugenic claims. It is now clear that

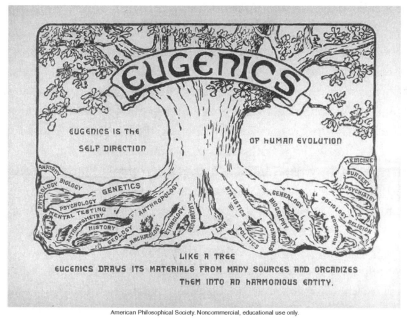

FIGURE 11-1. Tree logo of the American Eugenics Society.

even during the heyday of eugenics (1910–1935), there was no body of carefully collected and peer-reviewed data to back up its policies and the statements made by well-respected scientific "authorities." Data collected and reported by eugenics researchers suffered from many deficits—lack of clearly defined traits (feeblemindedness), bias in data collection (lower IQ test scores of non-English-speaking immigrants), and outright falsification of data. Even the data that had some objective validity—such as family histories of mental illness and the ethnic makeup of prison populations—did not take into account the impact of social and economic factors that might lead people from disadvantaged families or social groups to be imprisoned or diagnosed as mentally ill at greater rates than people from more economically privileged social groups.

These eugenics scientists, some of whom made meaningful contributions in other fields, did not apply rigorous scientific principles to the social aspects of genetics. There is also ample evidence that eugenics laws were implemented with little regard for justice or due process—falsified evidence, collusion between defense and prosecution attorneys, and bogus expert testimony were common in these cases. It seems that enthusiasm for eugenics as the social application of the modern science of genetics was so fueled by prejudices against unpopular racial and ethnic groups and disabled people that it overwhelmed the usual safeguards of both science and the law.

The history of scientifically unsound eugenic social policy is not confined to the early twentieth century. Laws against intermarriage between black and white people remained in force until the late 1990s in some U.S. states. Pseudo-genetic arguments for the intellectual inferiority of blacks (or other groups targeted by racists) compared to whites can still be found in public discourse—such as the 1994 book *The Bell Curve* by Herrnstein and Murray. Other simplistic explanations such as the existence of single genes responsible for complex human behaviors (violence, alcoholism, homosexuality) continue to surface in the scientific literature and the mainstream press. It is important to keep in mind that the evidence for simple genetic controls of intelligence, violence, criminality, or any other form of human behavior is no more convincing today than it was in the 1920s. But, as more human genes are mapped and their functions investigated, these simplistic explanations are likely to increase. Simplistic explanations are inevitable, as the media reduce complex scientific discoveries to 30-second items on the evening news: "Gene for alcoholism found; story at eleven." Thus there is a need for genomic scientists to pay special attention to the press coverage and to the social implications of their work.

In 1970, the federal government initiated a program of mandatory screening for nonsymptomatic carriers of the sickle cell anemia trait. This screening was primarily implemented among children entering the public school system. Some insurance companies denied coverage or charged higher premiums for people who carried this trait (overwhelmingly African Americans), even though possession of this trait carries no health risk (and, in fact, it provides improved resistance to malaria). Many African Americans believed that this screening program was the first step in a policy of genocide by preventing sickle cell carriers from marrying and having children.

This historical perspective may create an atmosphere of popular distrust surrounding the claims of currently respected scientists when they speak about the social applications of genomic technologies. If the now-absurd claims of eugenics were approved by the scientific and legal authorities of the early twentieth century, how can the public be sure that modern medical genomics technologies are not also influenced by the social and economic biases of today's society? What seems like objective, evidence-based medicine right now may appear laughably naive in a few generations.

HUMAN GENOME DIVERSITY PROJECT AND POPULATION GENETICS

One aspect of the HGP the Human Genome Diversity Project (HGDP), was designed to collect and analyze DNA samples from a wide range of ethnic groups and genetically distinct or isolated populations to gain a greater understanding of human genetic diversity and evolution. The history of the HGDP provides a modern example of ethics in action as applied to genomics technology. One major motivation for the HGDP was the belief among anthropologists and population geneticists that the HGP would sequence DNA samples from primarily Anglo-European people, thus neglecting important diversity in the world's other populations. The members of many minority groups and ethnically distinct populations met the proposal with a level of suspicion similar to the sickle cell screening program. They characterized the HGDP as the "Vampire Project," which would take their blood but return no benefits to them—a form of molecular colonialism. In fact, the prevailing belief among members of such groups was that any genetic information discovered about distinct populations would be used to discriminate against them.

Scientists involved in the HGDP were bewildered by the outpouring of negative feelings by members of indigenous groups, since their own work on the project was motivated by a genuine interest in the evolutionary history of humanity and a belief that knowledge of human genetic diversity would lead to medical benefits for all people, particularly the members of these groups. It is not that the HGDP lacked proper informed consent protocols for its sampling methods, but that all Western scientists lack credibility among many members of indigenous groups (Cunningham and Scharper, 1996).

While the HGDP was not directly related to any initiatives to patent genes or cell lines, the NIH was correctly identified as a major sponsor of the HGDP and as the applicant for a number of gene and cell line patents. In March 1995, the NIH obtained a patent on a cell line infected with a leukemia-associated virus from a man from the Hagahai people of Papua New Guinea (PNG). The NIH and the Centers for Disease Control and Prevention (CDC) also pursued patents on cell lines derived from blood donated by a woman from the Guaymi population of Panama and from a person from the Solomon Islands. These patent applications led to considerable negative publicity for the NIH and for the concept of genome prospecting by Western scientists among people from less-developed countries.

The NIH did not pursue these patents to make money from selling drugs or from contracts with drug companies. Rather, federal government policy encourages the NIH to transfer information, materials, and intellectual property rights to private companies who can develop and manufacture drugs and medical devices needed to care for the sick. In the current legal climate, this technology transfer can best be done if NIH holds the patent, so that a corporate partner can receive a clear and unencumbered right to develop a drug based on NIH discoveries. This approach makes sense from the perspective of U.S. science policy administrators, but it is counter to the feelings of many advocates for the rights of indigenous peoples. Ruth Liloqula, director of agricultural research for the Ministry of Agriculture and Fisheries of the Solomon Islands stated: "In our culture, genes are not inventions" (quoted in Kreeger, 1996). The underlying social contract in American society is that we all support biomedical research—through tax dollars, participation in clinical trials, or through our work as scientists and medical professionals—in return for improvements in health care. The reality in indigenous communities in developing countries is that the people will benefit much less than the average American from these improvements in health-care technology, especially protocols developed by large drug companies. In fact, opposition to the HGDP specifically and against population genetic research in general is quite widespread throughout the world. Examples of official declarations against the HGDP include the following: Karioca Declaration (June 1992, Brazil); Mataatua Declaration (June 1983, Aotearoa, New Zealand); United Nations Working Group on Indigenous Populations, 10th Session (July, 1993, Geneva); Maori Congress (1993, Aotearoa, New Zealand); National Congress of American Indians (December 1993, Resolution NV-93-118); Maori Congress of Indigenous Peoples Roundtable (June 1994); Guaymi General Congress (1994, Panama); Latin and South American Consultation on Indigenous Peoples Knowledge (September 1994, Santa Cruz de la Sierra, Bolivia); Asian Consultation on the Protection and Conservation of Indigenous Peoples Knowledge (February 1995, Sabbah, Malaysia), Pan American Health Organization Resolution (April 1995); Pacific Consultation on the Protection and Conservation of Indigenous Peoples Knowledge, Suva Statement (May 1995).

The NIH scientists involved in the PNG cell line project were working in good faith, following all available guidelines for the fair treatment of human subjects. In fact they made substantial efforts to be culturally sensitive. The project originated as part of a long-term study of the origin of human retroviruses throughout the world. NIH scientists collaborated with Carol Jenkins of the PNG Institute of Medical Research, who had been involved with many medical projects with the Hagahai people. Blood samples were collected by Jenkins from the Hagahai with the prior consent of local Hagahai leaders and the PNG government. Jenkins was listed as a co-inventor of the patent, and she pledged that her share of any patent royalties would be given to the Hagahai people.

Publicity surrounding the NIH patent resulted in a number of public accusations being made against Jenkins by various organizations working for the rights of indigenous peoples. The

Rural Advancement Foundation International (RAFI) claimed in a press release that "The United States Government has issued a patent on . . . an indigenous man of the Hagahai people" (Taubes, 1995). The PNG foreign affairs secretary later issued a statement exonerating Jenkins on behalf of the PNG government: "It is clear that this research has been done with the full consent of the Hagahai people as well as approval from the PNG Medical Research Institute and that the benefit of this research, when fully realized, will be shared among all concerned. " (PNG Press Release, 1996)

Despite the best intentions of the researchers, there is clearly an imbalance in the value proposition when genetic samples are collected from isolated populations. Communities of indigenous people are economically disadvantaged, and their way of life is under attack from a variety of social and environmental forces. By participating in a genetic research project, they give up a unique resource, yet receive little tangible value in return. Western scientists gain valuable information, which can be used to advance their personal research interests, their grant applications, and their careers. The information can then move on, via patents, publication, or inclusion in computer databases, so that it can be used by Western corporations for drug development research. The accusation is generally true that the researchers care more about the data than about the people from whom it is collected.

Groups that advocate for indigenous peoples argue that despite statements and rules to the contrary, the HGDP will be caught up in the current trend to commercialize genes. HGDP scientists, they fear, could simply become agents of the commercial interests of pharmaceutical companies. While a particular research project, such as the HGDP, may not be directly related to the patenting of a certain gene, it is clear that the net result of Western-government-sponsored genomics research will be commercial products, such as genetic tests and new drugs, and that large corporations will profit from these products—and populations of indigenous people will not. In fact, the specific information produced by the HGDP will be put into publicly accessible on-line databases, which will be much more valuable to international pharmaceutical corporations than it will be to the indigenous peoples from whom it was derived. In fact, it is entirely likely that some of the individual scientists involved in the HGDP have relationships with biotech corporations that lead to personal financial benefits or research sponsorship derived in some way from their work with samples from indigenous groups.

There are examples of DNA-based information obtained from isolated populations or indigenous peoples that have been commercialized for substantial profit. Scientists from Sequana Therapeutics (a California-based genomics company), collected samples from the people of Tristan de Cunha, a tiny island of just under 300 people located halfway between Brazil and South Africa. The inhabitants, who are all descendants of the island's original seven families, exhibit one of the world's highest incidences of asthma (30% of the population suffer from asthma and 20% are carriers). Sequana sold the licensing rights to a diagnostic test for asthma to the German pharmaceutical giant Boehringer Ingelheim for $70 million. Although the scientists and fieldworkers extracting the samples from indigenous populations follow informed consent guidelines, these guidelines often do not mention how their DNA or a product derived from it may become a marketable commodity that could potentially benefit private companies.

The appropriate remedy for these complaints of gene exploitation is both *obvious* and *impossible* under the current system of worldwide patents and intellectual property. The system allows patents for products of scientific work such as "discovered" natural chemicals, DNA sequences, and cell lines. This system has long benefited chemical prospectors who travel to remote locations, interview native people about their use of medicinal plants and other natural products, and then bring samples of those products back to corporate laboratories to be extracted and chemically characterized. Drugs and other chemical products discovered in the samples can then be sold by the companies without any compensation to the people from whom they were taken. It is clear that fear of the same practices underlie public outcry against the use of human samples from the indigenous peoples for genomic research.

The scientific members of the HGDP have created a rather noble (if practically unworkable) set of principles based on the concept of returning value to the indigenous people who contribute samples to the project: "The HGDP will not profit from the samples and it will do its best to make sure that financial profits, if any, return to the sampled populations." (North American Regional Committee, Human Genome Diversity Project, 1997)

Representatives of groups advocating for the rights of indigenous peoples have argued instead for a worldwide ban on the patenting of any living organism or DNA sequences derived from them. They argue that there is an inevitable connection between the patenting of life forms and the capitalistic expropriation of biological resources. The pharmaceutical industry would counter argue that there is a necessary link between patents and the investment in research that would develop a usable drug or DNA-based diagnostic product. Without patent protection, there can be no practical applications of medical genomics.

There is an important lesson to be learned here. Accusations of genetic exploitation were raised against scientists. After careful examination of the specific cases, all of the scientists involved were completely exonerated. Some additional changes were made in the laws, and charters were designed to make the actions of institutions such as the NIH and the HGDP more sensitive to the ethics of working with human genetic material. Yet the overall lasting impression in the public mind remains one of suspicion and distrust. The HGDP is still widely characterized as the Vampire Project. Accusations of exploitation are remembered by the public much more readily are than official denials of wrongdoing. Social activists have more public credibility on these issues than NIH scientists. Plans to educate the public to increase the acceptance of genomic technologies in medicine are likely to run into the same obstacles. During the PNG patent dispute, Temple University anthropologist and former NSF director Jonathan Friedlander noted a "widespread public distrust of the scientific technological enterprise and a willingness to believe the worst of people with scientific knowledge (Taubes, 1995)."

GENETIC DISCRIMINATION

One of the primary public concerns in America and Europe about genomic technologies—particularly about genetic testing—is the potential for genetic discrimination. This is discrimination against an individual or against members of that individual's family (or ethnic group) solely because of differences in DNA sequence that are not associated with any presently observable disease symptoms. It is based on the notion that people with certain genetic characteristics that put them at increased risk of disease will be charged higher insurance rates, denied certain jobs, etc. Most Americans believe that this information should not be available to employers and insurance companies. Fortunately, governments have been heeding this public sentiment and laws are being put into place to ensure the confidentiality of genetic information and to ban the use of genetic information in employment and health insurance. However, it will be difficult to build up public trust in these laws as adequate protection against genetic discrimination. Clearly, the physician will be called on to convince patients of the adequacy of the laws designed to prevent genetic discrimination when it becomes necessary to obtain consent for a genetic test.

One somewhat naive view being promoted by science policy advisory groups is that the public will accept new genetic technologies once they are better informed about the goals and benefits of the project and better educated about the underlying scientific principles. These scientists and public officials dramatically underestimate the fundamental distrust that many segments of the public hold for institutions such as the government, insurance companies, and the health-care system, when it comes to genetics. The crucial issue is not that the public does not understand the technology (although it is true that they do not), it is that they do not trust the sources of the information or the motives of the institutions sponsoring the projects.

The case for the social risks of genetic tests is not getting any easier, since a few companies will inevitably abuse the technology. In a recent court case, the Burlington Northern Santa Fe Railroad admitted to the use of genetic tests in an attempt to disprove claims for compensation from employees with carpal-tunnel syndrome. Employees who filed an injury claim were forced to provide a blood sample. The company submitted the samples for a genetic test for hereditary neuropathy without the consent of the employees. However, since carpal-tunnel syndrome does not have a genetic basis, the illegal testing had no basis in science—more likely it was an ill-conceived attempt to confuse the issue in court cases concerning the payment of workers' compensation. The forced testing was stopped by order of the federal Equal Employment Opportunity Commission (EEOC) in February 2001; and in March, the company apologized to the employees who were subjected to testing, agreed to destroy all blood samples and records of the test results, and to pay damages and legal fees. However the net result of this well-publicized case is not to instill faith in the vigilance of federal agencies but rather to confirm public suspicions that genetic tests will be made in secret and that there will be conspiracies to use the results against people.

Ironically, at the present time there seems to be little evidence that health insurance companies are using genetic information or that they plan to use such information in the future. Among health insurance and health maintenance organizations (HMOs), the prevailing opinion is that the results of genetic tests have too little predictive value about the short-term health prospects of a person. Most Americans change heath insurance plans quite frequently, so long-term health risks are not particularly important when making coverage decisions. Furthermore, the vast majority of Americans have health insurance plans provided by the government (Medicare and Medicaid) or from their employer, and such coverage is automatic, cannot be disallowed by current health status or preexisting conditions, and costs the same for everyone. The Federal Health Insurance Portability and Accountability Act (HIPAA) of 1996 has significantly diminished the threat of genetic discrimination in group health insurance. HIPAA forbids insurance plans to label otherwise healthy people who have a genetic condition as having a pre-existing condition.

There is considerably more interest in genetic information among life insurance providers. Many genetic factors can influence a person's life expectancy, and thus the profit or loss that might be expected from a life insurance policy purchased by that person. Insurers have claimed that knowledge of these factors would enable them to more accurately evaluate risk and more fairly set premiums for each person. However, it seems unlikely as a social policy, in America or Europe, that life insurance companies will be allowed to set premiums or deny coverage based on genetic tests. Insurers are concerned that individuals may discover, through genetic testing, that they have a substantial risk of contracting some serious condition and then decide to purchase a large amount of life insurance. If the insurer does not have access to the genetic information, then it could lose money on the policies purchased by individuals who know more about their own life expectancy than the insurance company. It is quite difficult to design an equitable social policy that can address this imbalance of information.

IMPACT ON THE PHYSICIAN AND RESEARCHER

Many of the ethical issues being raised in regard to medical genomics are not new: safety and efficacy of new treatments, privacy, discrimination, and informed consent. It has long been the case that many kinds of medical information (not just genetic) that resides in computer databases (or paper files) can be damaging to a person if they were to get into the wrong hands—an employer, an insurance company, or the press.

These are some of the key arguments that have been raised about genetic data, all of which also apply to other kinds of medical data:

It is predictive
- But so is testing for any kind of infection with a long latency for the development of symptoms, such as routine cancer screening, HIV, and sexually transmitted diseases (STD).

Privacy and confidentiality are important
- But obviously this applies for all medical information, including such routine information as family medical history and STD.

It involves risk factors and probability
- But ordinary testing for cholesterol level implies a risk for heart disease.

It has social, family, insurance, and discrimination impact
  • But so does psychiatric disease, AIDS, and STD.

However, there are important concerns about informed consent for medical genomics that do require special attention from medical professionals. This is particularly true when genetic tests can reveal potential risk factors that do not indicate that a person currently has a disease and in fact may not ever develop it; or when a test indicates the presence of a disease for which there is no treatment. In fact, it may be quite sensible for a person to choose not to have a genetic test that might predict a future disease for which there is no cure and no preventative measures. The key issue here revolves around providing enough information to the patient so that he or she can make a truly informed decision about whether to undergo a genetic test (informed consent). There is also a strong requirement for careful communication of test results to the patient and to family members, so that they can fully understand the implications and make informed health-care decisions.

The burden of communication placed on the physician in order to achieve informed consent is indeed a heavy one. Not only must patients be educated in the relevant molecular biology needed to understand the nature of a deleterious allele and the mechanics of the test by which it is identified, but they must also be led to understand the nature of risk and probability. There is a substantial body of social science studies showing that people tend to misinterpret statistical information and make demonstrably poor choices that work against their own interests (e.g., public lotteries and other games of chance). It is not at all clear that a physician (or genetic counselor) can ever provide enough information to patients to allow them to make truly informed choices for their own best interests in regard to a genetic test. Furthermore, no health-care professional ever enters a discussion with a patient without some subtle bias as to what course of action would be best for that patient. Again, the bias of the professional often strongly influences the decision of the client.

One concern that is particularly enhanced in the era of genetic testing is the persistence of genetic data and the ease by which many tests can be applied to a tiny sample and that genetic information gathered for one purpose can later be reanalyzed to reveal other information about a person. In the era of medical genomics, a routine blood or tissue sample taken for some innocuous purpose (such as a pharmacogenomic test before taking a drug) can provide a complete genetic profile of a person. Even if the sample is destroyed, the DNA sequence information obtained in one test, may reveal other sensitive information, for example, a gene form that causes susceptibility to a particular drug side effect may also be linked to a higher risk of some form of cancer. Clearly there is a need for strong laws backed up by well-executed policies and procedures to prevent unauthorized genetic testing of people and of access to their genetic information, wherever it may exist (in patient records, in computer databases, etc.).

This concern has been addressed by legislation at the state, federal and international levels. The Genetic Privacy Act was proposed as federal legislation in 1995, but never voted into law by Congress. President Clinton signed an executive order in February 2000, prohibiting federal agencies from using genetic information to discriminate against employees. The Genetic Nondiscrimination in Health Insurance and Employment Act (S 318/HR 602) was debated in both the Senate and the House of Representatives in 2001. The bill would prohibit insurers from rejecting anyone or adjusting fees on the basis of genetic information; make genetic discrimination in all areas of employment illegal, including hiring and compensation; and forbid insurers and employers from requiring genetic testing. The American Civil Liberties Union is supporting this law.

The Health Insurance Portability and Accountability Act, often referred to as HIPAA, was passed by Congress and signed into law by President Clinton in August 1996. Among many other provisions, the law prohibits health insurance plans from taking into consideration genetic information in determining eligibility for coverage or in setting premiums. Many of the law's key provisions will not be fully implemented until 2002–2003. Note that HIPAA applies only to group health insurance—not to private insurance purchased directly by an individual, disability, long-term care, or life insurance.

By the middle of 2001, a total of 37 states had enacted their own legislation to provide some forms of protection for the confidentiality and use of genetic test results. These state laws have prohibitions against genetic discrimination that extend to employment, certain commercial transactions, health and disability insurance, long-term care, and life insurance. These laws place heavy burdens on the physician to understand and implement these protections. In particular, physicians will need to do the following:

- Identify genetic tests and protected genetic information under the law's confidentiality provisions.
- Obtain the required written consents both to conduct a genetic test and to release the test results to anyone other than the patient
- The physician must provide information to the patient about the reliability of the test and the availability of follow-up genetic counseling.
- Educate office staff and institute office procedures to ensure appropriate handling of genetic information.

On learning of all of these new legal obligations, a physician might prefer not to engage in any genetic testing, just to avoid new hassles. However, it turns out that a substantial amount of information that is already in patient charts—such as family histories and enzyme tests—actually contains potentially protected genetic information under these new laws. Genetic testing in its simplest form takes place in doctors' offices, clinics, and hospitals every day. Talking to a health-care provider about your family history can reveal genetic information about your current health and predisposition to disease; and this information becomes part of your permanent medical record. According to Charles A. Welch, vice president of the Massachusetts Medical Society, "Since genetic information is ubiquitous in patient records, the requirement that physicians separate genetic information from the medical record is, in many cases, a requirement to do the impossible" (quoted in Green and Nicastro, 2001). The new laws being created as a result of public fears about the misuse of new genetic testing data may have the added benefit of improving the overall privacy of medical data.

Another ramification of new medical data privacy laws, especially HIPAA, is that biomedical researchers will have more difficulty obtaining research samples. Traditionally, any tissue that was removed from a patient during surgical procedures in a hospital was considered fair game for medical research studies. Often the samples were "anonymized" in some way so that there was no way of identifying the person who "donated" the sample. Many of the new medical privacy laws clearly specify that all tissue samples removed from a person are the personal property of that person, and the hospital can use those samples only in ways for which the person specifically provides informed consent. This complicates the research process in many ways. First, it is not always known beforehand what research studies will be conducted with a given sample. Tissue may be frozen or stored in other ways for future use, transferred to distant laboratories for a particular study; later, another study may be added to the project, etc. How can the patient consent in advance for a study that was not planned at the time the sample was collected? Alternately, it would be difficult to recontact every patient to obtain consent for some later study on stored samples. Also, if documents must be maintained proving that informed consent has been given for the use of each sample in each research study, then those documents themselves become a security risk, since the identity of the patient is linked to the sample.

It is obviously counterproductive for privacy laws to block the basic research that is needed to develop the genetic tests that provide the results that are the subject of this debate. There would be no debate over genetic testing if there were no clear benefits and widespread public demand for this testing. We all want the benefits of genetic screening for treatable diseases and drugs tailored to our genetic profile without the risks of genetic discrimination. There is no reason that carefully crafted laws cannot be created to reach this goal. However, the legacy of past unethical uses of genetics creates an emotionally polarized debate and deep-seated distrust of scientists; government; and drug, medical, and insurance companies around these issues.

Despite all of the recommendations of advisory committees and task forces and the adoption of statements of principles by scientific bodies, religious organizations, and governments, the public remains convinced that genetic information can and will be used against them. This is a well-founded fear—and one that is shared by many health-care professionals when it comes to their personal medical records. Clearly, medical professionals are going to face a significant barrier of public mistrust before the benefits of routine genetic testing can be realized.

REFERENCES

Cunningham H, Scharper S. Human Genome Project patenting indigenous third world. 2/23/96 http://www.dartmouth.edu/~cbbc/courses/bio4/bio4-1996/HumanGenome3rdWorld.html. Accessed 2/10/2002.

Davenport CB. Eugenics: The science of human improvement through better breeding New York: Henry Holt and Company, 1910.

ELSI Research Planning and Evaluation Group. A review and analysis of the Ethical, Legal, and Social Implications (ELSI) research programs at the National Institutes of Health and the Department of Energy: Final Report of the ELSI Research Planning and Evaluation Group. Feb. 10, 2000.

Galton F. Essays in eugenics. London: Eugenics Education Society, 1909.

Green MJ and Nicastro DP. New state genetic privacy act. Vital Signs (Massachusetts Medical Society) Feb. 2001; http://www2.mms.org/ vitalsigns/feb01/lr3.html.

Hernstein, R. and C. Murray. The Bell Curve: Intelligence and Class Structure in American Life. 1994. The Free Press, New York. 845 pp.

Image Archive on the American Eugenics Movement. vector.cshl.org/ eugenics.

Kreeger KY. Proposed human genome diversity project still plagued by controversy and questions. Scientist 1996;10(20):1.

Laughlin HH. Report of the Committee to Study and to Report on the Best Practical Means of Cutting off the Defective Germ Plasm in the American Population. Eugen Rec. Office Bull. 1914; 10:1–64.

Lombardo PA. Three generations, no imbeciles: New light on Buck v. Bell. NY U Law Rev 1985;60:30–62.

North American Regional Committee: Human Genome Diversity Project. Model Ethical Protocol for Collecting DNA Samples. Houston Law Review. 1997; 33(5): 1431–1473.

Papua New Guinea Secretary for Foreign Affairs and Trade. Press release, Mar. 1996.

Taubes G. Scientists attacked for patenting pacific tribe. Science 1995;270:1112.

GLOSSARY

α-helix

The most common 3-dimensonal secondary structure for polypeptide chains (proteins), determined by Linus Pauling in 1951. It looks like a spiral staircase in which the steps are formed by individual amino acids spaced at intervals of 1.5 Å, with 3.6 amino acids per turn. The helix is held together by hydrogen bonds between the carbonyl group (COOH) of one amino acid residue and the imino group (NH) of the residue 4 positions further down the chain.

accession number

A unique number assigned to a nucleotide, protein, structure, or genome record by a sequence database builder.

algorithm

A step-by-step method for solving a computational problem.

alignment

A one-to-one matching of two sequences so that each character in a pair of sequences is associated with a single character of the other sequence or with a gap. Alignments are often displayed as two rows with a third row in between indicating levels of similarity. For example:

```
GCT---GTCTGAACCCAACCAGACGGAGAATGA
: ::    ::: :: : :    :: :    ::::::  ::
GCTCCTGTCGGACCTCCTGCAGGGGGAGAACGA
```

allele

Alternate forms of a gene which occur at the same locus (see polymorphism). All of the variant forms of a gene that are found in a population.

alternative splicing

Variations in the process of removing introns from the primary transcript of a gene that lead to different mature mRNAs.

annotation

The descriptive text that accompanies a sequence in a database record.

anti-codon

The three bases of a tRNA molecule that form a complementary match to an mRNA codon and thus allow the tRNA to perform the key translation step in the process of information transfer from nucleic acid to protein.

assembly

The process of aligning and building a consensus (contig) from overlapping short sequence reads determined by DNA sequencing.

autosomes

Chromosomes which are not involved in sex determination.

β-Pleated Sheet

A protein secondary structure in which two or more extended polypeptide chains line up in parallel to form a planar array which is held together by inter-chain hydrogen bonds. The pleats are formed by the angles of bonds between amino acids in the polypeptide chains.

BAC (Bacterial Artificial Chromosome)

A cloning vector based on the naturally occurring F-factor plasmid from *E. coli* that can contain from 100,000 to over 300,000 bases of inserted DNA.

base pairs

Hydrogen bonded pairs of DNA nucleotides. Adenine always pairs with Thymidine and Guanine bonds with Cytosine (A-T and G-C base pairs).

bioinformatics

The use of computers for the acquisition, management, and analysis of biological information.

BLAST (Basic Local Alignment Search Tool)

A fast heuristic database similarity search tool developed by Altschul, Gish, Miller, Myers, and Lipman at the NCBI that allows the entire world to search query sequences against the GenBank database over the web. BLAST is able to detect relationships among sequences which share only isolated regions of similarity. BLAST software and source code is also available for UNIX computers for free from the NCBI. Variants of the BLAST program include blastn (DNA query vs. DNA database), blastp (protein query vs. protein database), blastx (translated DNA query vs. protein database), tblastn (protein query vs. translated DNA database), and tblastx (translated DNA query vs. translated DNA database).

Boolean search terms

The logical terms "AND," "OR," and "NOT" which are used to make database searches more precise.

bottleneck

A severe reduction in the number of individuals in a population, leading to a reduction in the genetic diversity of that population in later generations.

"Central Dogma" of Molecular Biology

DNA is transcribed into RNA which is translated into protein (proposed by Francis Crick in 1957).

cDNA

Complementary DNA—a piece of DNA copied *in vitro* from mRNA by a reverse transcriptase enzyme.

chiasma

The physical crossover point between pairs of homologous chromosomes in the process of recombination which can be observed during the diplotene and diakinetic stages of prophase 1 and during metaphase 1 of meiosis.

chimera

A hybrid, particularly a synthetic DNA molecule that is the result of ligation of DNA fragments that come from different organisms.

chromosome

A complete DNA molecule which carries a set of genes in a linear array. The basic unit of heredity.

class prediction

A diagnostic method which reliably categorizes a sample into one of a defined set of classes based on an assay (e.g. acute myeloid leukemia vs. normal).

cloning

The process of growing a group of genetically identical cells (or organisms) from a single ancestor. Also, the process of producing many identical copies of a segment of DNA or a gene using recombinant DNA technology.

cloning vector

A DNA construct such as a plasmid, modified viral genome, or artificial chromosome that can be used to carry a gene or fragment of DNA for purposes of cloning.

coding sequence

The portion of a gene that is transcribed into mRNA.

codon

A linear group of three nucleotides on a DNA segment that codes for one of the 20 amino acids (see genetic code).

conserved sequence

A base sequence in a DNA molecule (or an amino acid sequence in a protein) that has remained essentially unchanged throughout evolution.

contig

A consensus sequence generated from a set of overlapping sequence fragments that represent a large piece of DNA, usually a genomic region from a particular chromosome.

diploid

A genome (the DNA contained in each cell) that consists of two homologous copies of each chromosome.

divergence

The gradual acquisition of dissimilar characters by related organisms over time as two taxa move away from a common point of origin (see sequence divergence).

diversity

The number of base differences between two genomes divided by the genome size.

domain

A discrete portion of a protein with its own function. The combination of domains in a single protein determines its overall function.

dominant

An allele (or the trait encoded by that allele) which produces its characteristic phenotype when present in the heterozygous condition.

DNA

Deoxyribonucleic Acid, the information containing part of chromosomes that is responsible for both the transmission of hereditary traits and the moment by moment control of cellular physiology.

DNA Sequencing

The laboratory method of determining the nucleotide sequence of a piece of DNA, usually using the process of interrupted replication and gel electrophoresis developed by Fred Sanger.

EMBL (European Molecular Biology Laboratory)

The European branch of the 3 part International Nucleotide Sequence Database Collaboration (together with GenBank and DDBJ) which maintains the EMBL Data Library (a repository of all public DNA and protein sequence data). Each of the three groups collects a portion of the total sequence data reported worldwide, and all new and updated database entries are exchanged between the groups on a daily basis. However, database files obtained from EMBL are in a different format than those obtained from GenBank. The EMBL, established in 1974, is supported by 14 European countries and Israel. Like the NCBI, the EMBL also provides extensive bioinformatics tools.

enhancer

A regulatory DNA sequence that increases transcription of a gene. An enhancer can function in either orientation and it may be located up to several thousand base pairs upstream or downstream from the gene it regulates.

ENTREZ

Entrez is the online search and retrieval system that integrates information from databases at NCBI. These databases include nucleotide sequences, protein sequences, macromolecular structures, whole genomes, and MEDLINE, through PubMed.

EST

Expressed Sequence Tag—a partial sequence of a cDNA clone created by collecting single sequencing runs from the 3' and 5' ends of a cDNA clone.

e-score (Expect value)

The Expect value (E) is a parameter that describes the number of hits one can "expect" to see just by chance when searching a database of a particular size. An e-value of 1 is equivalent to a match that would occur by chance once in a search of that database.

exon

A segment of an interrupted gene (i.e. a gene that contains introns) that is represented in the mature mRNA product—the portions of an mRNA that is left after all introns are spliced out, which serves as a template for protein synthesis.

FASTA

A fast heuristic sequence similarity search program developed by Pearson and Lipman. Searches for local regions of similarity between sequences, tolerant of gaps. The related programs TFASTA compares a protein query sequence to a DNA databank translated in all six reading frames and TFASTX compares a protein sequence to a DNA database taking frame-shifts into account.

FASTA format

A simple universal text format for storing DNA and protein sequences. The sequence begins with a ">" character followed by a single-line description (or header), followed by lines of sequence data.

founder effect

Differences in the allele frequencies of a specific sub-population as compared with the rest of the species due to random differences in the small number of alleles carried by the individuals who were the founders of the sub-population.

functional genomics

The study of the function of every gene and protein in the genome including roles in metabolism, physiology, development, regulatory networks, etc.

gap

A space inserted into a sequence to improve its alignment with another sequence.

gap creation penalty

The cost of inserting a new gap in a sequence when creating an alignment and calculating its score.

gap extension penalty

The cost of extending an existing gap by one residue in an alignment.

GenBank

A repository of all public DNA and protein sequence data. GenBank is the U.S. branch of the 3 part International Nucleotide Sequence Database Collaboration (together with EMBL and DDBJ). GenBank is currently administered by the National Center for Biotechnology Information, National Library of Medicine, Bethesda, Maryland, a division of the US National Institutes of Health.

gene

A segment of DNA sequence (a locus on a chromosome) that is involved in producing a protein. It includes regions that precede and follow the coding region as well as all introns and exons. The exact boundaries of a gene are often ill-defined since many promoter and enhancer regions dispersed over many kilobases may influence transcription.

gene expression

The process by which a gene provides the information for the synthesis of protein—i.e. transcription into mRNA followed by translation into protein.

gene expression profile

A pattern of changes in the expression of a specific set of genes that is characteristic of a particular disease or treatment (e.g. cancerous vs. normal cells). The detection of this pattern may be limited to a particular type of gene expression measurement technology.

gene family

A group of closely related genes that make similar protein products.

gene regulatory network

A map of the relationships between a number of different genes and gene products (proteins), regulatory molecules, etc. that define the regulatory response of a cell with respect to a particular physiological function.

genetic code

The correspondence between 3 base DNA codons and amino acids that directs the translation of mRNA into protein. There is one "standard" genetic code for all eukaryotes, but some prokaryotes and sub-cellular organelles use variant codes.

genetic determinism

The unsubstantiated theory that genetic factors determine a person's health, behavior, intelligence, or other complex attributes.

genetic engineering (*see recombinant DNA*)

genome

All of the genetic material in a cell or an organism.

Genome Ontology (GO)

A standard set of consistent naming conventions that can be used to describe gene and protein functions in all organisms based on molecular function, biological process, and cellular location.

genome project

The research and technology development effort aimed at mapping and sequencing the entire genome of human beings and other organisms.

genomics

The use of high throughput molecular biology technologies to study large numbers of genes and gene products all at once in whole cells, whole tissues, or whole organisms.

GenPept

A comprehensive protein database that contains all of the translated coding regions of GenBank sequences.

global alignment

A complete end-to-end alignment of two sequences. This can often be misleading if the two sequences are of different length or only share a limited region of similarity.

haplotype

A specific set of linked alleles from a group of adjacent genes that are inherited together over a number of generations.

helix-turn-helix

A protein secondary structure found in many DNA binding proteins. Two adjacent α-helixes are oriented at right angles to each other.

heterozygous

An organism (or cell) with two different alleles for a particular gene.

heterozygosity

The presence of different alleles of a gene in one individual or in a population. A measure of genetic diversity.

heuristic

A computational method based on a process of successive approximations. Heuristic methods are much faster, but may miss some solutions to a problem that would be found using more laborious rigorous computational methods.

HMM (Hidden Markov Model)

A statistical model of the consensus sequence of a sequence family (i.e., protein domain). HMMs are based on probability theory—they are "trained" using a set of sequences that are known to be part of a family (a multiple alignment), then can be applied on a large scale to search databases for other members of the family.

homology

Similarity between two sequences due to their evolution from a common ancestor.

homologs

Sequences that are similar due to their evolution from a common ancestor.

homozygous

An organism (or cell) with two identical copies of the same allele for a particular gene.

HSP (high scoring segment pair)

An alignment of two sequence regions where no gaps have been inserted and with a similarity score higher than a threshold value.

identity (*see sequence identity*)

informatics

The study of the application of computer and statistical techniques to the management of information. In genome projects, informatics includes the development of methods to search databases quickly, to analyze DNA sequence information, and to predict protein sequence and structure from DNA sequence data.

intron (intervening sequence)

A segment of DNA that is transcribed, but removed from the mRNA by a splicing reaction before translation into protein occurs.

in vitro

(Latin) Literally "in glass," meaning outside of the organism in the laboratory. Usually in a tissue culture.

in vivo

(Latin) Literally "in life," meaning within a living organism.

Ligase

An enzyme which can use ATP to create phosphate bonds between the ends of two DNA fragments, effectively splicing two DNA molecules into one.

linkage

A relationship between two genes located nearby on a single chromosome where the combination of alleles found in each parent appear together in the progeny more frequently than would be expected by chance.

linkage analysis

The process of locating genes on the chromosome by measuring recombination rates between phenotypic and genetic markers (or finding markers that do not recombine away from a phenotype).

linkage disequilibrium

A set of alleles that remain more tightly linked than would be expected by chance among the members of a population.

locus

A specific spot on a chromosome—the location of a gene, a mutation, or other genetic marker. A given locus can be found on any pair of homologous chromosomes.

MEDLINE (PubMed)

The U.S. National Library of Medicine's bibliographic database covering the fields of medicine, nursing, dentistry, veterinary medicine, and the biological sciences. The MEDLINE file contains bibliographic citations and author abstracts from approximately 3,900 current biomedical journals published in the United States and 70 foreign countries. PubMed is a web-based search tool for MEDLINE.

meiosis

The process of double cell division in a reproductive cell which produces haploid gametes.

microsatellite

A form of repetitive or low complexity DNA that is composed of a short sequence (1–15 bp in length) that is tandemly repeated many times. This is often a hotspot for mutations.

minisatellites

Repetitive DNA sequence composed of tandemly repeating units of 10–100 base pairs.

mismatch

In an alignment, two corresponding symbols that are not the same.

mitosis

The process of cell division which produces a pair of daughter cells that are genetically identical to each other and to the parent cell.

motif

A region within a group of related protein or DNA sequences that is evolutionarily conserved—presumably due to its functional importance.

mRNA (messenger RNA)

RNA molecules that are synthesized from a DNA template in the nucleus (a gene) and transported to ribosomes in the cytoplasm where they serve as a template for the synthesis of protein (translation).

multiple alignment

The alignment of three or more sequences—usually done by the progressive pairwise method—which yields an approximate rather than an optimal answer.

mutation

A change in DNA sequence.

neutral mutations

A change in DNA sequence which has no phenotypic effect (or has no effect on fitness).

NCBI (National Center for Biotechnology Information)

A branch of the U.S. National Library of Medicine, which is part of the NIH. The NCBI is the home of GenBank, BLAST, MedLine/PubMed, and ENTREZ.

non-coding sequence

A region of DNA that is not translated into protein. Some non-coding sequences are regulatory portions of genes, others may serve structural purposes (telomeres, centromeres), others have no known function.

OMIM (Online Mendelian Inheritance in Man)

An online database of human genes and genetic disorders authored and edited by Dr. Victor A. McKusick. The database contains textual information, pictures, and reference information. It also contains copious links to NCBI's Entrez database of MEDLINE articles and sequence information.

ORF (open reading frame)

A region of DNA that begins with a translation "start" codon (ATG) and continues until a "stop" codon is reached—this usually understood to imply a protein coding region of DNA or an exon.

orthologs

Similar genes or proteins (homologs) that perform identical functions in different species—identical genes from different species.

paralogs

Similar genes or proteins (homologs) that perform different (but related) functions either within a species or in different species—members of a gene family. The line between orthologs and paralogs grows less distinct when proteins are compared between distantly related organisms—is a bacterial protein an ortholog of a human protein which performs an identical function if the two share only 15% sequence identity?

PCR (Polymerase Chain Reaction)

A method of repeatedly copying segments of DNA using short oligonucleotide primers (10–30 bases long) and heat stable polymerase enzymes in a cycle of heating and cooling so as to produce an exponential increase in the number of target fragments.

Pfam

An online database of protein families, multiple sequence alignments, and Hidden Markov Models covering many common protein domains, created by Sonnhammer ELL, Eddy SR, Birney E, Bateman A, and Durbin R. Pfam is a semi-automatic protein family database, which aims to be comprehensive as well as accurate.

pharmacogenomics

The use of associations between the effects of drugs and genetic markers to develop genetic tests that can be used to fine-tune patient diagnosis and treatment.

phylogenetics

Field of biology that deals with the relationships between organisms. It includes the discovery of these relationships, and the study of the causes behind this pattern.

phylogeny

The evolutionary history of an organism as it is traced back connecting through shared ancestors to lineages of other organisms.

plasmid

A circular DNA molecule that can autonomously replicate within a host cell (usually a bacteria).

polymorphism

A difference in DNA sequence at a particular locus.

position-specific scoring matrix

A table of amino acid frequencies at each position in a sequence calculated from a multiple alignment of similar sequences.

post-transcriptional regulation

Regulation of gene expression that acts on the mRNA (i.e. after transcription). This includes regulation of alternative intron splicing, poly-adenylation, 5′ capping, mRNA stability, and rates of translation.

post-translational regulation

Regulation of gene expression that acts at the protein level (i.e. after translation). This includes differential rates of protein degradation, intracellular localization and/or excretion, internal crosslinking, protease cleavage, the formation of dimers or multi-protein complexes, phosphorylation and other biochemical modifications.

primer

A short DNA (or RNA) fragment which can anneal to a single-stranded template DNA to form a starting point for DNA polymerase to extend a new DNA strand complementary to the template, forming a duplex DNA molecule.

profile analysis

A similarity search method based on an alignment of several conserved sequences, such as a protein motif. The frequency of each amino acid is computed for each position in the alignment, then this matrix of position specific scores is used to search a database.

progressive pairwise alignment
A multiple alignment algorithm that first ranks a set of sequences by their overall similarity, then aligns the two most similar, creates a consensus sequence, aligns the consensus with the next sequence, makes a new consensus, and repeats until all of the sequences are aligned.

ProDom
An online protein domain database created by an automatic compilation of homologous domains from all known protein sequences (SWISS-PROT + TREMBL + TREMBL updates) using recursive PSI-BLAST searches.

promoter
A region of DNA that extends 150–300 base pairs upstream from the transcription start site of a gene that contains binding sites for RNA polymerase and regulatory DNA binding proteins.

ProSite
PROSITE the most authoritative database of protein families and domains. It consists of biologically significant sites, patterns and profiles, complied by expert biologists. Created and maintained by Amos Bairoch and colleagues at the Swiss Institute of Bioinformatics.

protein family
Most proteins can be grouped, on the basis of similarities in their sequences, into a limited number of families. Proteins or protein domains belonging to a particular family generally share functional attributes and are derived from a common ancestor.

proteome

All of the proteins present in a cell or tissue (or organism).

proteomics

The simultaneous investigation of all of the proteins in a cell or organism.

PubMed

PubMed is a web-based search tool for MEDLINE at the NCBI website.

query

A word or number used as the basis for a database search.

recombination

The crossing over of alleles between homologous chromosome pairs during meiosis which allows for new (non-parental) combinations of alleles to appear among genes on the same chromosome.

recombinant DNA cloning

The use of molecular biology techniques such as restriction enzymes, ligation, and cloning to transfer genes among organisms (also known as genetic engineering).

recessive

An allele (or the trait encoded by that allele) which does not produce its characteristic phenotype when present in the heterozygous condition. The recessive phenotype is hidden in the F1 generation, but emerges in 1/4 of the progeny from an F2 self-cross. Most genetic diseases are the result of gene defects which are present as recessive traits at low to moderate frequencies in the population, but emerge in progeny when two parents both carry the same recessive allele.

replication
The process of synthesizing new DNA by copying an existing strand, using it as a template for the addition of complementary bases, catalyzed by a DNA polymerase enzyme.

restriction enzyme
A protein, manufactured by a species of bacteria, which recognizes a specific short DNA sequence within a long double stranded DNA molecule, and cuts both strands of the DNA at that spot.

scoring matrix (substitution matrix)
A table which assigns a value to every possible amino acid (or nucleotide) pair. This table is used when calculating alignment scores.

segregation
The separation of chromosomes (and the alleles they carry) during meiosis. Alleles on different chromosomes segregate randomly among the gametes (and the progeny).

sequence identity
The percentage of residues identical ($D \rightarrow E = 0$; $D \rightarrow D = 1$) between two aligned sequences.

sequence similarity
The percentage of amino acid residues similar between two aligned protein sequences. Usually calculated by setting a threshold score from a scoring matrix to distinguish similar from not similar and counting the percentage of residues that are above this threshold.

SNPs
Single Nucleotide Polymorphisms; single base pair mutations that appear at frequencies above 1% in the population.

shotgun method

A sequencing method which involves randomly sequencing tiny cloned pieces of the genome, with no foreknowledge of where on a chromosome the piece originally came from.

signal sequence

A 16–30 amino acid sequence located at the amino terminal (N-terminal) end of a secreted polypeptide, that serves as a routing label to direct the protein to the appropriate sub-cellular compartment. The signal sequence is removed during post-translational processing.

significance

A statistical term used to define the likelihood of a particular result being produced by chance. Significance values for sequence similarity searches are expressed as probabilities (p-values or e-values) so that value of 0.05 represents one chance in twenty that a given result is due to chance.

similarity (*see sequence similarity*)

sister chromatids

A pair of homologous chromosomes aligned during meiosis.

Smith–Waterman algorithm

A rigorous dynamic programming method for deriving the optimal local alignment between the best matching regions of two sequences. It can be used to compare a single sequence to all of the sequences in an entire database to determine the best matches, but this is a very slow (but sensitive) method of similarity searching.

somatic

All of the cells in the body which are not gametes (sex cells).

structural proteomics

The study of 3 dimensional protein structures on all proteins in a cell, tissue, or organism as a guide to gene/protein function.

SwissProt

A curated protein sequence database which provides a high level of annotations, a minimal level of redundancy and high level of integration with other databases. SwissProt contains only those protein sequences which have been experimentally verified in some way—none of these "hypothetical proteins" of unknown function.

synteny

A large group of genes that appear in the same order on the chromosomes of two different species.

systematics

The process of classification of organisms into a formal hierarchical system of groups (taxa). This is done through a process of reconstructing a single phylogenetic tree for all forms of life which uncovers the historical pattern of events that led to the current distribution and diversity of life.

taxa

A named group of related organisms identified by systematics.

threading

A method of computing the 3-dimensional structure of a protein from its sequence by comparison with a homologous protein of known structure.

transcription

Synthesis of RNA on a DNA template by RNA polymerase enzyme.

transcription factor

A protein which binds DNA at specific sequences and regulates the transcription of specific genes.

transduction

The transfer of new DNA into a cell by a virus (and stable integration into the cell's genome).

transfection

The process of inserting new DNA into a eukaryotic cell (and stable integration into the cell's genome).

transformation

The introduction of foreign DNA into a cell and expression of genes from the introduced DNA (does not necessarily include stable integration into the host cell genome).

translation

Synthesis of protein on an mRNA template by the ribosome complex.

TrEMBL (translations of EMBL)

A database supplement to SWISS-PROT that contains all the translations of EMBL nucleotide sequence entries not yet integrated into SWISS-PROT.

UniGene

An online database (at NCBI) of clustered GenBank and EST sequences for human, mouse and rat. Each UniGene cluster contains sequences that represent a unique gene, as well as related information such as the tissue types in which the gene has been expressed and map location.

Index